Negotiating Climate Change

About the Author

Amanda Machin is a visiting lecturer at
the University of Westminster.

Negotiating Climate Change

Radical Democracy and the Illusion of Consensus

Amanda Machin

Zed Books
London & New York

Negotiating Climate Change: Radical Democracy and the Illusion of Consensus was first published in 2013 by Zed Books Ltd, 7 Cynthia Street, London N1 9JF, UK and Room 400, 175 Fifth Avenue, New York, NY 10010, USA

www.zedbooks.co.uk

Designed and typeset in Goudy and Eurostile by Kate Kirkwood
Index by John Barker
Cover designed by www.kikamiller.com
Printed and bound by
CPI Group (UK) Ltd, Croydon, CR0 4YY

Distributed in the USA exclusively by Palgrave Macmillan, a division of St Martin's Press, LLC, 175 Fifth Avenue, New York, NY 10010, USA

A catalogue record for this book is available from the British Library
Library of Congress Cataloging in Publication Data available

ISBN 978 1 78032 398 5 hb
ISBN 978 1 78032 397 8 pb

For Kester, Olive, Isla and Hazel

Contents

Acknowledgements

The research from which this book arose was carried out while I held a post-doctoral research position with the Governance and Sustainability programme at the University of Westminster. I am very grateful to Simon Joss for his support and leadership of this programme. I am also indebted to the colleagues and friends who took the time and trouble to read through and comment on draft chapters and previous versions of the arguments found in this book: Dan Greenwood, Paulina Tambakaki, Bronwyn Hayward, Maria Fotou and Chantal Mouffe. I am also grateful to Tamsine O'Riordan and Kim Walker for their invaluable editorial guidance and to two anonymous reviewers.

Writing a book having recently had a baby has been a rather daunting task, and I wholeheartedly thank my family who granted me the time and space in which to write and forgave me my fretting and fussing. Thank you to Brian, Tricia, Pat and Les, and most of all, of course, to Paul.

Introduction
Where are the Politics of Climate Change?

Climate change, it has been said, is the greatest challenge of our time.[1] This is true in two senses. It appears, first, as a global issue, rearing up as a cataclysmic rupture facing the entire planet and thus shaping the existence of present and future generations for decades to come. The second challenge is that it seems to be a huge problem with no straightforward and forthcoming solution. It is thus an apparition that looms and lurks in our everyday lives, a shadow that falls across our tomorrows and haunts our consciences. This shadow is all the more disturbing because we do not seem to be able to work out a strategy for how it might be banished.

Might this apparition be rendered less fearsome if we notice that climate change is the silhouette of our own self-image – that it is indeed our own shadow? Might it help if we turned to look hard at ourselves and our ways of life and our values to see why we have created this thing and what it reveals about us? Until we do so, I suggest, we will not be able to dispel the all too real fear and concern that climate change is causing us.

This book, then, attempts to look at climate change by looking at ourselves. It suggests that the way we envisage ourselves as individuals and as collectivities affects our ability to respond to the varied impact and meaning of climate change. Perhaps by understanding ourselves differently we might be able to work out how to tackle the greatest challenge of our time. Perhaps a transformed self-awareness may help to render it not such an impossible challenge after all.

I shall argue that dominant approaches to climate change do not help us approach this multifaceted issue in any useful way; rather, they only reaffirm our predicaments. I attempt here to point to the flaws in the dominant perspectives and to draw out the common threads between them. One such thread is the problematic construction of climate change as *one* problem in need of one global solution, a solution that this construct itself inhibits. Another is the reification of a particular depiction of the individual, as self-interested and rational. A third strand is the assumption that we need a common understanding, reached through rationality, to act on climate change. I upend this assumption and claim that to have any hope of acting against climate change we need to disagree. I argue that the disagreement that Mike Hulme notices in his important book can be put to work in a radical democracy.[2]

The myth of consensus, I suggest, is perhaps the biggest problem facing climate change politics today. Consensus – both scientific and political – is supposed to underpin decisive action. But such consensus is not forthcoming. Scientists may be able to predict what might happen when atmospheric temperatures rise, although they cannot forecast with any accuracy the how and when and to whom, nor can they tell us what to do with this information. Shifting political decisions onto science results in a depoliticisation of climate change that dampens democratic debate and stifles political engagement. Assuming political consensus as a horizon marginalises those who dissent and undermines the role of disagreement in politics. Disagreement, I suggest, should be celebrated and encouraged in a democratic forum. For only through disagreement can climate change politics be invigorated and produce a clear decision, although this decision is never final and full. Such a radical democratic approach, I suggest, has more chance of producing the collective action to tackle climate change than current dominant approaches.

At present climate change is approached from very different angles that reveal particular self-understandings and desires. Some scour science and technology for a magical solution that allows (some) human beings to continue to live their lives without making any noticeable sacrifices. Yet the potential side-effects of such far-reaching, and sometimes far-fetched, mechanisms must necessarily be heeded – the clouds about the silver lining. The search for technology fits into a broader picture of market-based solutions, where economic prizes and punishments provide the motivation for environmentally friendly behaviour and products. In this picture, investors are prodded with subsidies to encourage them to develop new environmentally friendly technologies, industries are manipulated with emissions allowances, and the general population are coaxed away from their carbon-dependent lifestyles through green taxes. Chapter 1 examines the 'techno-economic' approach and explains that the policies it advocates cannot, by themselves, produce any real structural change; they are often invisible and demand no awareness or engagement of the dangers of climate change by the population (take away the tax or the emissions cap, it seems, and behaviour will revert). These sorts of policies can be criticised for actually affirming the very self-interested behaviour that is at issue. Techno-economic solutions are not strong, deep or long-lasting enough to tackle the problem, but need to be part of a clear and decisive policy framework.

Chapter 2 examines the 'ethical-individual' approach, which suggests that it is not the market or technology that proffers the solution but rather an invigoration or reconceptualisation of our ethics. Moral philosophers demand a transformation in fundamental ethical values to produce the massive transformations in human behaviour and attitudes needed to address climate change. This approach, however, again focuses upon individuals and their personal choices. The problem here is the assumption that individuals, acting independently, can agree on how to tackle climate change and will accordingly act harmoniously, without any political policy to coordinate and coerce them. Although climate change is indeed an ethical issue it is not only an ethical issue. An approach that focuses entirely on ethics is hindered by the difficulty that no decisive and collective action is guaranteed by a transformation of individual ethical values.

To underpin a move to a greener society in general, including more climate-friendly lifestyles, it seems that an explicit commitment to action as a collective is needed. The next chapters examine perspectives that crucially regard individuals as citizens. Chapter 3 analyses the 'green republican' approach which acknowledges the importance of community in underpinning a green way of life.

Green republicans argue that a commonly held understanding of 'the good' can unify a community, and thus they promote the existence of a common good which contains a substantive green component. Individuals here are not regarded as self-interested consumers but rather as responsible and virtuous citizens. The problem with green republicanism, however, is that it presupposes an agreement upon the substantive content of the 'common good' and citizenship responsibilities. Who decides the meaning and demands of the 'common good'? Does this approach result in an 'eco-authoritarianism' that dictates the interests, values and actions of a population? What, moreover, of the gender and class identifications that may result in varied understandings of and relationships to climate change? Green republicanism appears to demand an agreement that might well be precluded by pluralism and democratic differences.

Can we rely on our politicians to instigate and follow through on the requisite decisive action? Democratic publics don't look likely to provide the push to tackle climate change. For politicians to implement the necessary policies, they have to have the sanction of the people and many are sceptical that such sanction exists. Theorists and policy makers of climate change, therefore, often shy away from democracy, either by quietly undermining democratic publics, or like 'eco-authoritarians', by dismissing democratic politics completely. The 'deliberative democratic approach' considered in Chapter 4 seems to hold the key to overcoming the reluctance of a population in tackling climate change. Advocates of this approach promote a different, rejuvenated, form of democracy over conventional forms of representative democracy. They suggest that by being brought into contact with each other in a deliberative forum, apparently conflictual perspectives can be transformed through reasoned discussion that aims at a horizon of democratic consensus. This approach is concerned with overcoming disagreement, which it regards as ultimately an obstacle to collective action. But can disagreement be expunged? The deliberative democratic approach assumes that the political realm is a space in which discussion can be completely free, open and inclusive, and in which rational argument can address differences. I argue that this approach is misguided. Both *whether* and *how* we act to combat climate change is a difficult political choice on which there will always be disagreement.

Consensus on how to combat climate change cannot and will not ever be reached; there is no one 'rational' path to take, no overarching grand green scheme that suits everyone. Any apparently inclusive agreement and rational discussion is rather a trick of power that disguises exclusion and inequality. Rather than excluding differences, and attempting to eradicate disagreement, I suggest that differences should be both revealed and revered, for democratic differences are actually necessary to tackle the problem of climate change. Disagreement should not be regarded as an obstacle to decisive and collective action on climate change, but rather as a necessary requirement in achieving it. In Chapter 5 I present a 'radical democratic' approach. I suggest that the presupposition that consensus on an issue such as climate change can be reached is misguided. Flipping the dominant idea on its head, I argue that it is not agreement that should be searched for, but disagreement. Decisive action is underpinned not by consensus but by disagreement, for without a choice between real alternatives there can be no decision. It is disagreement that actually underpins decisive action; to make a political decision, it is not just that there *will* be disagreement but that there *must*. This disagreement, however, should be contained within democratic structures that ensure it will not erupt into violence and hostility. As I will show, following Chantal Mouffe's theory of agonism, this involves a 'conflictual consensus' in which democratic values are upheld by all, although there is no agreement on their substantive content.

In Chapter 6 I explain how a recognition of nature and climate as political categories can underpin democratic disagreement. Contrary to the assumption that fostering environmental awareness will produce a clear and unified set of ecological values, I argue that bringing the environment into our consciousness splinters perspectives. Rather than being just an *object* of disagreement, the environment becomes an important *ground* of disagreement. By focusing on how our various identifications are bound up with our environment we can therefore revitalise democratic politics. I suggest that we should shift attention away from global agreements and towards local disagreements.

The book, then, progresses towards the tentative presentation of radical democracy as an approach that could both give to and gain from the issue of climate change. I hope, however, that the chapters can be read as stand-alone arguments and in a different order to how they are laid out. I must point out here what this book does not do. First, I am not claiming to have covered all the different perspectives on climate change in this book, nor to have managed satisfactorily to divide up the rich body of theory on this topic into unified and distinct categories. Categorisation is always a particular work of construction that can never be definitive. I am aware that I have omitted important arguments and that I have squashed many into pigeon holes where they may not feel comfortable. Second, this is a book of political theory, and although it touches upon other disciplines I approach them from a political and theoretical perspective. It therefore lacks practical policy recommendations and economic or scientific detail. Nevertheless, I hope that this presentation of the issue of climate change, and the approaches taken in assessing it, provides a useful new stance – one from which to examine climate change by examining the possibility of change in ourselves.

1 Magic and Markets
The Techno-Economic Approach

Certainty and Climate Change

The wisest of you men is he who has realised, like Socrates, that in respect of wisdom he is really worthless – Plato, *The Apology*[1]

The Oracle of Delphi proclaimed Socrates to be the wisest man in Greece, to which Socrates responded that, if this was so, it was only because he alone was aware of his own ignorance. This response of Socrates contains something of the prophetic itself; it is a telling comment in the modern world, where ignorance is brushed off with fingertips that have at their immediate disposal reams of virtual facts and streams of direct data. In today's world, so similar and yet so different to that of Socrates, knowledge must be instant, succinct and assured. For a long time uncertainty was regarded as the basic condition of human existence, a lack of knowledge of the future that was assuaged only by the activities of gods, oracles and fortune tellers. Today this situation has been wholly reversed, and we expect to walk and talk upon surer grounds. Certainty is demanded and scientists are compelled to provide it.

Like the diviners of an earlier era, scientists and experts are understood to work and think in a loftier place, and are thus at once esteemed and suspected for harbouring higher knowledge. Yet science is very different from the forecasts of oracles and seers, for it works with a systematic rigour and upon the assumption that scientific claims are falsifiable. The practice of peer review and the constant drive to question distinguish science from avowals of faith.[2] Scientific statements about climate change, then, should be taken seriously.

What are scientists really saying and how certain can they be about saying it? In the early nineteenth century, the French physicist and mathematician Jean Baptiste Joseph Fourier noticed that the earth's atmosphere retained heat from the sun, making the planet liveable for humans. By the end of that century, a Swedish scientist called Svante Arrhenius had linked carbon dioxide (CO_2) to this process.[3] It is now understood that CO_2 as well as certain other gases (ozone, methane and water vapour) absorb some of the infrared radiation rising from the surface of the earth to keep the temperature of the planet warmer than it would be otherwise. These gases are known as greenhouse gases and, as their quantities rise, the 'greenhouse' becomes hotter and hotter. Further investigation discovered that the concentration of greenhouse gases in the atmosphere was growing, and that this was linked to human activity. The burning of fossil fuels (coal, oil and gas) for energy releases captured carbon into its gas form, and the simultaneous occurrence of deforestation around the world limits the capturing of this increased volume through photosynthesis.

The growth of greenhouse gases is exponential. In the twentieth century global human energy consumption has increased over 16 times.[4] Recent figures show that 180 new coal-fired power stations are built a year, over half of them in China (at a rate of two a week). A modern coal-fired power station can produce over 10 million tonnes of CO_2 per year.[5] Atmospheric CO_2 has risen from 275 to 370 parts per million (ppm).[6] Even if the production of greenhouse gases could be massively reduced, however, this would not prevent further global heating since it is the *accumulated* total of gases that determines temperature levels. The delicate mechanism that maintains the temperature of the earth within narrow limits is being destroyed.

To those who study climate change, and to a growing percentage of the general public also, this information is all too familiar. But despite an awareness of this evidence, the deficit of adequate action to tackle the problem it illustrates is well documented.[7] This apparent apathy might be partly attributed to the lack of any watertight scientific prediction. There is no definitive picture of what will happen to the world if greenhouse gases are generated at the same rate, and what will happen if they aren't.

The Intergovernmental Panel on Climate Change (IPCC) is an organisation set up in 1988 by the United Nations Environment Programme (UNEP) and the World Meteorological Organisation (WMO). Its widely used reports are written by 500 lead authors and 2,000 expert reviewers from over 100 countries. The latest IPCC report tells us that a warming of the earth's climate is now 'unequivocal' and that the observed change is 'very likely' due to human generated greenhouse gas emissions.[8] The likely outcome is an increase in heat waves, cyclone intensity, flooding in some areas, water stress in others, erosion, loss of biodiversity and a damaging impact upon human health, especially to populations with 'a low-adaptive capacity'. The International Food Policy Research Institute (IFPRI) claims that climate change could lead to avalanches, erosion, tsunamis and volcanic eruptions, precipitating massive migration, devastating food shortages and global conflict.[9] But none of this is certain. It cannot be. Science cannot accurately forecast the future regarding climate change; there are too many variables, too many unknowns, too many 'ifs', for it to do that. Scientists themselves tell us that they cannot tell us this. They readily acknowledge that they cannot give us exact figures and precise scenarios. As the IFPRI report states, climate change simulations are 'inherently uncertain'.[10]

In a path-breaking account of the quality of scientific epistemic claims in today's world, Silvio Funtowicz and Jerome Ravetz suggest that climate change is an example of an issue that needs a different type of science – a 'post-normal science': 'Whereas science was previously understood as steadily advancing in the certainty of our knowledge and control of the natural world, now science is seen as coping with many uncertainties in policy issues of risk and the environment.'[11] Science, they explain, has led us into problems that it cannot solve, by grossly manipulating the natural world and by oppressing alternative forms of knowledge. There are today an abundance of issues for which facts are uncertain, values are disputed and decisions are urgent. They reject the term 'problem' for such issues, since this term implies that there is an expectation of a clear solution. Normal science struggles to consider such issues and instead a post-normal science is needed in which 'uncertainty is not banished but is managed'.[12] They suggest that responding to these issues requires an extension of the decision-making community beyond the scientific elites to new participants. Local knowledge of particular facts and unofficial information can enrich the scientific investigation: 'Knowledge of local conditions may determine which data are strong and relevant, and can also help to define the policy problems.'[13]

The assumption that scientific uncertainty can and must be removed has problematic normative and political implications.[14] Some accounts of climate change, juxtaposing science and policy, suggest that the obstacles to tackling climate change are political rather than scientific – implying that the problems lie entirely in responding appropriately to the science through political institutions. For example, Steve Vanderheiden argues that 'anthropocentric climate change involves highly complex causal chains and is expected to produce consequences that are extraordinarily difficult to forecast, yet perhaps the most confounding aspects of the problem are political rather than scientific'.[15] He might be right, but I would formulate this situation differently. Politics and science cannot so easily be separated out: it is the very difficulty in forecasting the changes in temperature, weather, sea levels and so on that makes the problem a political one, and it is the unpredictability of human activity and politics that heightens the difficulty of scientific predictions.

It is not just that scientific knowledge about the future will always be incomplete and uncertain, but that scientific knowledge itself is always shaped by social processes.[16] As Mike Hulme writes, 'where science is practised, by whom and in what era, affects the knowledge that science produces'.[17] Hulme is highlighting the important insight that individuals – whether scientists or artists or high court judges – cannot step outside their social reality to 'know' the world objectively. As human beings we are embedded within the social world we are born into and no 'transcendental knowledge' is possible. Hulme warns against policy makers just accepting the scientific consensus of the day, and argues that the process of establishing scientific knowledge should remain transparent and open. Scientists should aim to communicate the uncertainty about climate change rather than expunging it: 'far from being able to eliminate uncertainty, science – especially climate change science – is most useful when it finds good ways of recognising, managing and communicating uncertainty'.[18]

Even when fairly certain of itself, however, science cannot tell us what to do with its predictions; scientific assessments do not contain any implicit policy guidance. We can examine the data and draw conclusions, but we cannot 'read off' the meaning of these measurements; they do not, in themselves, tell us what to do. The IPCC explains that 'determining what constitutes "dangerous anthropocentric interference with the climate system" … involves value judgements'.[19] Yet policy makers seem both to demand certainty and to assert and assume that it already exists. In this mode the UK Climate Change Committee announced a target: 'The world needs to aim to limit the temperature increases to 2 degrees Celsius … we must act now, and act globally.'[20] This palpable urgency is to be welcomed, perhaps, but the figure of 2 degrees Celsius is not given; it has been chosen as an ambitious but achievable policy objective. Other targets could have been selected. James S. Risbey explains that science cannot tell us what the thresholds of dangerous climate change should be and urges us to 'acknowledge at the outset the arbitrary and conditional nature of any specific choice or definitions of what is dangerous and what is not'.[21]

Scientific knowledge about an issue such as climate change can only ever be partial. As Sheila Jasanoff puts it, 'science cannot tell us where and when disaster will strike, how to allocate resources between prevention and mitigation, which activities to target first in reducing greenhouse gases, or whom to hold responsible for protecting the poor'.[22] For Roger A. Pielke, 'Science never compels just one political outcome. The world is not that simple.'[23] We cannot hide behind science when ethical and political choices must be made: too often science is a front, obscuring the decisions that are actually taking place. 'In current political decision making, scientific prognoses … act as "fig leaves" that hide the actual decision-making process and the normative assumptions on which it rests.'[24] As David Demeritt explains, the conventional view of science as 'hermetically sealed off' from politics is hugely naïve. 'Not only has the science of climate change largely driven the national and international politics of climate change, the politics in turn have also influenced the practice of that science.'[25] The point here is not to discredit climate science but to become more sensitive to its social construction. Once this is observed, moreover, one sees how the notion of one universal solution undermines the different implications that can be drawn from scientific data by different regions and communities.

Science gives us a picture of the future, but it is a sketchy one, its lines drawn in thick brush strokes depicting a climate that continues to warm, along with rising seas, flooding and erosion. But science cannot fill in the details of this picture. It can tell us what might happen, but not what will happen, and least of all what should happen. Rather than heaving the political responsibilities of climate change onto scientists, rather than demanding certainty when no such certainty exists, we need to see climate change science as a socially and politically enmeshed discipline that cannot present indubitable facts about the future. It should be regarded as an invaluable toolkit to be used in different contexts as necessary rather than an omniscient oracle that provides the truth. Following Socrates on cleverness-in-ignorance, perhaps we should note the limits of our knowledge – and acknowledge, work with and perhaps even celebrate uncertainty.

What I am calling the 'techno-economic approach' to climate change contains a tendency to expect and demand science to act as seer; science is not expected to make the choices but rather to render such choice unnecessary. Political decisions are rendered inevitable; choice is expunged. But not only is science expected to tell us what to do, it is also pressed for how to do it. Not only is the problem supposed to be made clear and unproblematic, but so are the solutions. In the next section I examine the problematic assumption that science and technology can, and will, provide the instruments to solve the problems of climate change. Here, again, politics hunkers down behind science in a wholly inadequate response to climate change.

Technological Magic

For many years those who worried about climate change and those who worried about energy security were on opposite ends of the debate. It was said that we faced a choice between protecting the environment and producing enough energy. Today we know better. These challenges share a common solution: technology. By developing new low-emission technologies, we can meet the growing demand for energy and at the same time reduce air pollution and greenhouse gas emissions. As a result, our nations have an opportunity to leave the debates of the past behind, and reach a consensus on the way forward. And that's our purpose today.
George W. Bush [27]

In a speech in 2007, US President George W. Bush stated the intention to tackle climate change and energy security through the development of clean energy alternatives: nuclear, solar and wind power. In this statement, Bush promises that the answer to climate change is to be found through the design and implementation of new technology. In 2010 US President Barack Obama makes a very similar claim linking nuclear power to energy security and climate change: 'To meet our growing energy needs and prevent the worst consequences of climate change, we'll need to increase our supply of nuclear power. It's that simple.'[28]

This approach fits within the dominant understanding of climate change, in which technological solutions are expected to appear inevitably if and when the heating of the atmosphere through human activity becomes desperate. The right technology, this story suggests, will allow us to live the lifestyles that some have become accustomed to and to which others aspire. Renewable energy from wind, solar thermal, photovoltaic, geothermal, tidal and hydropower sources, nuclear fission and nuclear fusion, biofuels, and carbon sequestration are all championed as our possible technological saviours from the dystopian future of environmental catastrophe.

As examined above, science is expected to produce clear and indisputable predictions. But, in an extension of this mistaken picture of science, it is also prodded and probed for potential solutions, too. As Garrett Hardin wrote nearly 50 years ago: 'An implicit and almost universal assumption of discussions published in professional and semi-popular scientific journals is that the problem under discussion has a technical solution.'[28] A technical solution, he goes on to explain, 'may be defined as one that requires a change only in the techniques of natural sciences, demanding little or nothing in the way of change in human values or ideas of morality'.[29] Hardin was writing about the population problem, which, he explained, belonged to a class of problems to which there simply are no technical solutions. He suggested that the problems of overpopulation can only be combated by enforcing coercive policy to curb the 'freedom to breed'. We don't have to agree with his conclusion to see the point he is making, which is that some problems cannot be solved through technical fixes but demand social and political change. His further observation is that the reason technical solutions are so appealing is that they avoid any need to alter our comfortable ways of living: 'most people who anguish over the population problem are trying to find a way to avoid the evils of over-population without relinquishing any of the privileges they now enjoy'.[30] This corresponds to the calls today for a 'technological revolution' in response to climate change. For if we can develop and implement the right technology, then surely we can sustain the existence of our cake while we continue to eat it.

It is not just the US government that searches for such magical solutions. In the UK, the government's recent 'Carbon Plan', published on the Department of Energy and Climate Change (DECC) website, contains similar claims, asserting that the UK can fulfil its pledge to cut greenhouse gas emissions and move to a low-carbon future through technological developments that will keep energy prices down and create jobs. 'This plan shows that the UK can move to a sustainable low carbon economy without sacrificing living standards, but by investing in new cars, power stations and buildings.'[31] On the Department for Environment, Food and Rural Affairs (DEFRA) website, in among the dry reports and wordy reviews, is a telling document entitled 'Future world images'. This document consists of six pictures of the world in the year 2030, illustrating some 'good adaptations' to climate change. Depicted are a domestic house, a farm, a city-scape, the countryside, the coast, and a scene of major infrastructure. Each image is labelled with measures to mitigate the negative effects of climate change and to exploit the opportunities; measures such as sustainable drainage, grey water, wind farms, carbon capture, flood plains. The document is careful to note that the images are not supposed to be providing definitive solutions but instead are 'thought provoking'; it states that 'they do not attempt to provide definite answers or solutions as the most appropriate action will depend on local circumstances'.[32] The images, then, are not concrete predictions of the world in 2030, but nor are they prescriptions. Do they 'set us thinking' as intended? Not, I would argue, in any radical way. The images look very familiar; the houses and farms and motorways in them are the houses and farms and motorways that exist now, with new technological fixes added on like sticking plasters.

What is shown by these images is a world of unchanged lifestyles. Motorways, office buildings and shops look a little different but remain central features of our current way of life. The lone hiker in the countryside has been 'educated' about how to avoid damaging it, but is not hiking any less, or anywhere different. In this story tackling climate change is a matter of developing and implementing new technologies, or re-educating ourselves to alter our lifestyles, but not in any detrimental or radical way.

But where does such technology come from? It often appears as if technological fixes emerge smoothly and logically from laboratories and factories. Yet the scientists who develop and discuss the technologies point out the importance of political will and leadership.[33] They suggest that a straightforward technological solution is highly unlikely, and certainly nowhere near entering a pipeline. They acknowledge, too, the political decisions that must be made in implementing technologies. The Royal Society, in its assessment of the potentials of geo-engineering, notes that while this 'deliberative large-scale intervention in the climate system' holds many scientific difficulties and uncertainties, these are dwarfed by the challenges of governing it: 'The greatest challenges to the successful deployment of geo-engineering may be the social, ethical, legal and political issues associated with governance, rather than scientific and technical issues.'[34] The use of geo-engineering raises crucial ethical and political questions: 'Far more detailed study would be needed before any method could even be seriously considered for deployment on the requisite international scale. Moreover, it is already clear than none offers a "silver bullet".[35] For example, some methods of geo-engineering can actually *reduce* levels of greenhouse gases in the atmosphere, generating the question of what the 'right' level of atmospheric concentration might be. Who makes such decisions? Do certain states or individuals get to decide the optimum composition of the global atmosphere?[36]

I am not arguing here that scientific technological innovation has no place in a response to global climate change. And I am not arguing that we should discard the hope that a combination of different technologies can stabilise and then reduce the atmospheric concentration of greenhouse gases. I am merely noting that we cannot assume that perfect technological solutions will arrive smoothly at our carbon-addicted fingertips as soon as we need them, and I am concerned by the propensity to make this assumption. Why is such a presupposition made? In the next section I suggest that the technological approach fits within a broader economic picture – one that regards the market as a mechanism through which an invisible hand works more or less by itself, both rationally and efficiently, to solve human problems. The production of the 'right' technology is just one important operation of the market.

Market Mechanisms

The technological solutions that promise magical solutions to climate change don't just augur the continuation of Western consumer lifestyles, they also offer increased employment and investment opportunities. This is the ideal vision of 'ecological modernisation': a low-carbon world, where energy is plentiful, clean and cheap, and the economy is booming.[37] The question of how climate-friendly technology comes about is often not asked, since the expectation is that it will simply arise through the supply-and-demand logic of the market. The promise of technology overlaps with a broader picture in which action to combat climate change comes about 'rationally' through creative entrepreneurs exploiting opportunities to make a profit. Here the market is the mechanism that can ultimately tackle climate change, which is regarded as predominantly an economic issue.

This approach emphasises the large amount of investment required to fund the development and implementation costs of the right technology. Advocates of such a market-based approach disagree about how much the market needs to be guided by policy. Some portrayals of 'ecological modernisation' suggest that business and government work hand in hand.[38] But, with or without government help, the climate is saved because it is good for business. If the short-term costs are high, it is assured, in the medium-to-long term there are economic benefits for all. The situation is ultimately 'win-win'; both the environment and ourselves, it seems, will become better off through innovation of climate-friendly apparatus. In his introduction to the 'Carbon Plan', Chris Huhne, secretary of state for energy and climate change, pointed the way forward: 'In June 2011, the Coalition Government enshrined in law a new commitment to halve greenhouse gas emissions, on 1990 levels, by the mid-2020s. This Carbon Plan sets out how we will meet this goal in a way that protects consumer bills and helps to attract new investment in low carbon infrastructure, industries and jobs.'[39] A further statement on the DECC website explains that the development of renewable energy sources, and carbon capture and storage, is central to the UK's efforts in tackling climate change: 'Our drive to increase the proportion of energy we obtain from renewable sources will not only increase the security of energy supplies in the UK; it will also provide opportunities for investment in new industries and new technologies.'[40] In referring to the Copenhagen Climate Summit, the then British prime minister, Gordon Brown, explained: 'The UN talks are ... not only about safeguarding the environment but also about stimulating economic demand and investment.'[41] And the current prime minister, David Cameron, sums up the 'win-win' idea of ecological modernisation in a speech on the green economy: 'deliberate investment in renewable energy isn't just good for our environment. It's good for our business too.'[42]

The assumption of a solution that benefits both economy and the environment was supported by the publication of the 2006 Stern Review, a widely cited report on climate change commissioned by the British government and put together by a team headed by Sir Nicholas Stern. The review offers a cost–benefit analysis suggesting that the economic benefits in tackling climate change outweigh the costs. It calculates that the cost of not acting will be equivalent to 5 per cent of global GDP each year, compared to the cost of action at 1 per cent of global GDP each year: 'the benefits of strong and early action far outweigh the economic costs of not acting'.[43] It asserts the importance of international frameworks of action, but argues that developing countries would not be impeded in their economic growth: 'action on climate change is required across all countries, and it need not cap the aspirations for growth of rich or poor countries'.[44]

The calculations in the review have been challenged.[45] But what is hardly ever criticised is the review's reduction of climate change to economics. The Stern Review is of course an economic analysis, so to criticise it for its stated purpose would be unfair. Economic analyses are, moreover, an important component in any overall account of climate change. But the review doesn't offer any acknowledgement that it provides only one perspective upon the issue, that there may be other sorts of values and ideas that should be taken into account in understanding climate change. It focuses entirely upon the economic consequences of climate change and the policies it advocates in countering climate change are also articulated entirely in economic terms. One of its key recommendations, for example, is that deforestation should be prioritised over reducing the carbon emissions of the transport sector because this is more economically viable: 'curbing deforestation is a highly cost-effective way to reduce emissions'.[46] But is such a policy politically, morally or institutionally sound? What are the social consequences of targeting deforestation in the parts of the developing world where it mainly occurs? Is it ethically justifiable to focus upon the actions of the developing world rather than the transport sector of the developed world? Other recommendations in the review include increased funding and cooperation in developing technology, increased funding for developing countries, and the expansion and interlinking of emissions trading schemes. These solutions have implicit, and questionable, presuppositions about the priority of economics.

And it is important to note that economists themselves, who agree with the actual economic focus of the Stern Review, have disagreed with its conclusions. William D. Nordhaus has challenged its 'time discount rate' – the relative weight assigned to the welfare of future generations. The Stern Review uses a 'zero time discount rate' which means that it treats future generations symmetrically with current generations. Nordhaus disputes this assumption, explaining that the assigning of discount rates have always been debated by economists and philosophers. He points out that 'the [Stern] Review should be read primarily as a document that is political in nature and has advocacy as its purpose'[47] and he concludes that 'the central questions about global-warming policy – how much, how fast and how costly – remain open'.[48] Whether or not Stern's discount rate is correct is an ethical question, which I will not consider here. But the important point is that the economic analysis cannot provide one undisputed 'rational' policy. Major ethical and political questions have been overlooked, as if the economic analysis had no need of them.

As a follow-up to the review, Nicholas Stern published a book entitled *Blueprint for a Safer Planet*. In it he typifies again the techno-economic approach, assuming the 'win-win' scenario in which the demands of the economy and the environment correspond: 'We have to see the issues of economic development and of climate change as parts of a whole.'[49] He explains that 'We can create a new era of progress and prosperity. We will discover new technologies and sources of energy along the way and make our energy supplies more secure. Ultimately, we will create a path of growth over the medium and long term, and generate major new opportunities for jobs and industry.'[50] In this approach it seems that climate change should simply be regarded as a business opportunity and the environment as a commodity with only economic value, as a report by the World Business Council for Sustainable Development (WBCSD) makes clear: 'Crisis. Opportunity.... The perfect storm we face, of environment, population, resources and economy will bring with it many opportunities.'[51] Is it advisable to regard climate change as a 'perfect storm' for business opportunity? Can it be reduced to that?

The other, related, assumption found in the techno-economic

approach is that the primary motivation for acting against climate change is economic. It is presupposed that if it can be shown that it is in everyone's economic self-interest to tackle climate change, it is just a matter of revealing this, and the necessary behaviour will result. The supply and demand logics of the market are supposed to guide every individual's behaviour in an economically rational way. The market is supposed to work its magic to tackle climate change with just a little prodding.

Although Stern, for example, notes that government intervention is key to tackling climate change, his book nevertheless regards the market and market enterprise as the principle mechanisms of a shift to a greener world. 'If public actions and policies give the right signals and rewards for cutting greenhouse gases, then markets and entrepreneurship will drive the response ... it is about enabling markets and private sector initiative to work well.'[52] He believes that prices should be 'at the heart' of policy, since 'the right prices can give an incentive for consumers and producers to do something'.[53] The market also ensures, he claims, that reductions in greenhouse gases are achieved as efficiently as possible. Hence, he argues for a combination of policies of taxation, trading and regulation.

The presuppositions of this approach are nicely summarised in the words of Richard Branson, founder and head of Virgin Group and arguably the very personification of the free market: 'At the moment the tone of voice around sustainability implies sacrifice and giving stuff up ... the opposite needs to be true.'[54] Can the difficulty in tackling climate change really be reduced to the resistance of the individual to material sacrifice? Is it just a matter of appealing to economic self-interest? Branson assumes that an environmentally friendly lifestyle will only appeal to people if it contains an ongoing supply of 'stuff'. Is he correct?

Taxing and Trading

The techno-economic approach is arguably the dominant approach to climate change today. In line with the prioritisation of the economic is the focus by the British government upon economic policies in attempting to encourage more environmentally friendly behaviour. Economic policies include green taxes, emissions trading, congestion charges and financial subsidies for the research and development of 'green' technologies. These policies assume that individuals are motivated, largely and primarily, by financial penalties and material rewards.

A major debate dominating environmental policy at present consists of assessing the relative merits of carbon taxes and emissions trading. Carbon taxes are perhaps the most straightforward and comprehensible tool for many states to use in their attempts to reduce carbon emissions of their populations. Taxes are put on either the emissions of 'upstream' producers or 'downstream' consumers, and place a fixed price on carbon. They are expected to provide strong incentives to reduce emissions, and they generate revenue that can be reinvested in other environmental policy measures or used to lower other taxes.[55] As Nordhaus puts it, 'it is critical to have a harmonised carbon tax or the equivalent both to provide incentives to individual firms and households and to stimulate research and development in low-carbon technologies'.[56] Nordhaus advocates carbon taxes over emissions trading as a flexible and efficient mechanism to limit the damage of global warming.[57] Yet there are problems with carbon taxes. They tend to have a disproportionate effect on low-income groups.[58] And they also penalise those who have high energy needs through no fault of their own: people, for example, who live in places exposed to extreme weather conditions or in areas that are 'structurally disadvantaged' through their lack of advanced energy-saving technologies.[59] Many advocates of green taxes are sensitive to these problems, and recognise that they would need to be implemented alongside other policies.[60]

An alternative policy tool is emissions trading. There are various schemes, the two most important being the Clean Development Mechanism (CDM) and the European Union's Emissions Trading Scheme (EU ETS). The CDM is an 'offsetting' scheme, established in 2003 by the Kyoto Protocol; it works by allowing developed countries to offset their emissions by helping to finance carbon-saving practices in developing countries.[61] For example, if a country classed as developing (so-called Non-Annex 1 countries) builds an energy-efficient gas power station instead of one powered by more polluting coal, the difference in potential carbon emissions is converted into CDM credits and can then be sold to a country classed as fully developed (so-called Annex 1 countries). These credits can be used by the Annex 1 country to count as part of its own emission reductions agreed in the Kyoto Protocol. The EU ETS is a 'cap and trade' scheme that divides out a limited number of emissions permits which can then be traded. The scheme is regarded as the 'cornerstone' of the EU's environmental policy and central to its strategy to meet its Kyoto target – to cut its greenhouse gas emissions to 8 per cent below 1990 levels by 2008–2012 – and also to establish the EU Commission as a leader in climate change policy.[62]

Emissions trading is supposed to be a more economically efficient policy than carbon taxing, but it too has problems.[63] As Fabian Schuppert explains, one of its main weaknesses is also regarded as one of its strengths. It operates in a way that doesn't harm the economy: 'existing cap-and-trade schemes are not radical enough, as the reason why they are promoted is precisely that they are economy friendly, which means that emission reductions are low, allocation of permits is mainly free ... and overall environmental impact is minimal'.[64]

However, while policy makers focus upon the debate over taxes and trading, as if it were a choice between one or the other,[65] they ignore the weakness of both types of policy. Such schemes are problematic in their assumption that climate change can be tackled by putting a 'price tag' on carbon or energy. The increased price, as we have seen, is supposed to incentivise individuals, groups and nations to lower their carbon consumption. But there are other motivations influencing energy users, apart from economic ones. For example the CDM trading scheme is unlikely, by itself, to persuade developing countries to implement more environmentally friendly ways, if that means a greater reliance on foreign imports.[66]

Oscar Reyes and Tamra Gilbertson offer a strong critique of carbon trading, suggesting that it precludes any proper discussion about the major necessary changes required for making a society more environmentally friendly. They claim that such a scheme 'closes down the space for asking the very questions that are crucial if we are to make the structural changes that might tackle climate change'.[67] There is no room in such an approach for questions of social justice, and of the long-term strategy for social change. Carbon trading, they claim, actually ends up increasing emissions, and they suggest that they simply distract us from the task: 'Ultimately, carbon trading is a means of pre-empting and delaying the structural changes necessary to address climate change.'[68] The 'structural changes' demanded by Gilbertson and Reyes are greater public investment, more accountable public institutions, and the move away from the promotion of grand schemes towards a 'multitude of locally adapted technologies and practices'.[69]

I believe that their criticisms apply equally to carbon taxes. A problem with all market-based policy instruments is that – by themselves – they don't lead to the requisite change in the social structure. After all, as Scott Prudham points out, the 'relentless, restless and growth-dependent character of capitalism's distinct metabolism' must surely clash with any attempt to move to a sustainable future.[70] Prudham regards announcements from entrepreneurs promoting a shift to 'green capitalism' as merely performances that work to legitimise the status quo. Certainly, there is little that is revolutionary about schemes such as the CDM and EU ETS, which are far removed from people's everyday concerns, while carbon taxes are often 'invisible' to consumers. These policies don't draw people's attention to the issue of climate change; there is no discussion of, let alone preoccupation with, the ways of life and behaviour patterns that must change if climate policies are going to work. As Nordhaus acknowledges, when trying to tackle an issue like global warming it is vital, as he puts it, 'to reach through governments to the multitude of firms and consumers who make the vast number of decisions that affect the ultimate outcomes'.[71] Yet the harmonised carbon tax that he advocates would fail to do this. As Tina Fawcett and Yael Parag explain, 'most existing market-based instruments do not seek to engage the public.... Greater engagement of citizens may be a necessary condition for delivering the systemic change required to achieve a low-carbon society.'[72]

These sorts of economic policy may have a role to play in tackling climate change. But they can only work efficiently if they are implemented alongside other policies and in the context of a decision, explicitly shared with citizens, to tackle climate change. Despite the fact that scientists and businesses demand that governments should take the lead, governments shift their responsibility back to science and markets.[73] Many of the advocates of market-based approaches seem to marginalise the role of political commitment, assuming either that innovation and new markets will flourish without government subsidy and support, or that such green policies would alienate populations who are more worried about the economic crisis than the ecological one. But this presupposes a certain conception of the human individual. Perhaps the most important problem with a narrow focus upon economic policy is that it conceives human beings reductively, in terms only of economic self-interest; individuals are seen primarily as consumers. In the next section I suggest that this reduction is inaccurate and that it fails to instigate a real shift to a climate-friendly society.

Rational Reductions

Dale Jamieson is one of many theorists who question the 'techno-economic approach' and its implicit assumption that human beings are motivated solely by economic self-interest: 'People often act in ways that are contrary to what we might predict on narrowly economic grounds, and moreover, they sometimes believe that it would be wrong or inappropriate even to take economic considerations into account.'[74] For Jamieson, neither scientific nor economic analysis is enough. Neither can tell us what we should value and how we should act: 'science has alerted us to a problem, but the problem also concerns our values'[75] and 'economics may be able to tell us how to reach our goals efficiently, but it cannot tell us what our goals should be or even whether we should be concerned to reach them efficiently'.[76] Jamieson queries Richard Branson's notion that the driving motivation behind human action is inevitably and unchangeably the quest for more and more 'stuff'. For Jamieson, what he calls 'management approaches' aimed at manipulating behaviour by providing economic incentives through taxes and subsidies and so on, are bound to fail, for they have an erroneous understanding of what motivates human beings.

One danger in the appeal to economic self-interest is that it can become a self-fulfilling prophecy; in introducing economic measures such as a carbon tax, the issue of climate change might be reframed as something that *is* simply a matter of money. Economic policies whitewash over the other values and questions around climate change, and the environment becomes another commodity. As Andrew Dobson explains, measures such as these reinforce an attitude that is part of the problem. He argues that rather than the encouragement of private individual economic self-interest, what is needed is the nurturing of a more public-spirited frame of mind: 'Taxes, fines, exemptions, rewards and permits all point away from the public towards the private, which is precisely the wrong direction.'[77]

Research suggests that policies designed to change behaviour through economic self-interest, far from being the solution, may actually be part of the problem. In a famous piece of research on blood donation, Richard Titmuss discovered that when people were paid to give blood rather than asked to give it voluntarily, they actually gave less. Titmuss argued that the commercialisation of blood – through which blood became a commodity rather than a gift – changed the norms associated with blood donation; it became a matter of economics rather than ethics. His conclusion was damning: 'the commercialisation of blood and donor relationships represses the expression of altruism, erodes the sense of community, lowers scientific standards, limits both personal and professional freedoms … increases the danger of unethical behaviour in various sectors of medical science and practice....'[78] He asks: 'Where are the lines to be drawn … if human blood be legitimated as a consumption good?... All policy would become in the end economic policy, and the only values that would count are those that can be measured in terms of money....'[79]

The philosopher Michael Sandel has interpreted Titmuss's findings as an example of 'market norms eroding or crowding out non-market norms'.[80] Sandel observes that leaving decisions to the market invites us to think of ourselves as consumers rather than citizens, and he is doubtful that this will produce the social transformation necessary to tackle climate change: 'If the countries of the world are able to change patterns of energy use and bring about a meaningful reduction in greenhouse gas emissions, it will not be because emissions trading schemes allow countries to buy and sell the right to pollute. Market mechanisms can be useful instruments. But real change will depend on changing people's attitudes toward nature, and rethinking our responsibilities toward the planet we share.'[81] The concern, then, is that by paying for carbon, people feel that they are entitled to it, and that any further obligation to reduce their carbon consumption disappears. Once we have paid our dues, we don't need to consider the impact of our lifestyles.

These sorts of mechanisms may simply reinforce the practices of consumption; economic 'solutions' work with the existing structure rather than challenging it. Unless these policies are part of a larger and wider structural change, they can easily be reversed. One of the apparent strengths of taxes is that they can be modified easily.[82] But this means that they can be removed easily. If environmental taxes and allowances are unpopular, as they are often expected to be, removing them might be seen as a quick way for a government to gain popularity. Any enduring green transformation of society therefore needs something less modifiable; it needs a firmer commitment and a shift to an entirely new set of values.

As Sandel explains, dominant socio-political conceptions today regard individuals as atomistic 'unencumbered selves' who are detached from their social environments and identities. These individuals are supposed to be rational and self-serving actors who behave in predictable ways that prioritise their own interests, understood in terms of economic well-being. But this model has been challenged by recent studies that undermine the notion of apparently rational, unencumbered and disembodied minds.[83] And such weirdly rational and atomised individuals would surely be debilitatingly disorientated in their understanding of themselves and the world: 'To imagine a person incapable of constitutive attachments such as these is not to conceive an ideally free and rational agent, but to imagine a person wholly without character, without moral depth.'[84] The policies of taxing and trading considered above presuppose that individuals are motivated only by economic self-interest and thus actually reproduce this attitude, an attitude that only hampers the social changes necessary to tackle an issue such as climate change. As Wynne puts it, 'mainstream social science tends to reinforce an atomised and instrumental rational choice self-interest model of the human subject'.[85] These theorists all recognise that a more useful picture of the human individual reveals them to be entwined within their social existence, encumbered and coloured by their political responsibilities, ethical ideals and collective identifications. This alternative picture of the individual is not only more convincing but might be crucial to underpinning a shift to a different type of society, in engendering an adequate response to issues such as climate change.

To summarise, then, I do not argue here that new technologies and market mechanisms do not have a place in achieving a low-carbon society. What I do suggest is that they cannot, by themselves, magically produce solutions. Implementing a science-based or market-based policy must involve a real engagement with the people who are affected by it. To make a real, lasting change, there must be an engagement with the issue of climate change by a greater proportion of the public. Without some sort of explicit commitment to climate-friendly living, there won't be any substantive transformation of the social structure. The problem, of course, is that it is the absence of this sort of commitment that is generally lamented. How can such a collective commitment be generated? I suggest that it is precisely through democratic disagreement in the political realm that citizens become inspired, engaged and interested in an issue. By shrugging off responsibility onto the economic realm, rather than seeing climate change as an issue of politics, the opportunity for engagement and lasting structural change is lost. The sorts of policies and perspectives advocated by the techno-economic approach sometimes actually reinforce the attitudes that are part of the problem. I will return to this argument after examining the second dominant approach in tackling climate change: the ethical-individual approach.

2 Good Consciences
The Ethical-Individual Approach

'Make Love not Carbon'

Be ethical, we are told. Be green. Recycle your waste, take public transport, change your light bulbs. Climate change is infiltrating our everyday life through the imperatives of green ethics that holler from the lifestyle pages of magazines, from the mouths of the celebrities, and from the printed slogans of reusable bags that are *not* plastic bags. These individual gestures, presumably, are supposed to amass into a colossal coordinated movement to counter climate change. But can they really do that? What do these eco-ethical imperatives tell us about ourselves?

The approach to climate change that I examine in this chapter is one that regards it as a symptom of a deeply unethical tendency in our current ways of living. Here climate change is an ethical matter, something that requires a transformation in the norms and values of human societies. Advocates of this sort of perspective suggest, quite correctly, that technological innovation and market mechanisms do not constitute, by themselves, an adequate response to climate change, and that a more radical structural shift is demanded. They comprehend both the problem of climate change and its solution through an ethical framework; it is a transformation not of our economics and technology but of our ethics that is demanded. Climate change is not a problem to be managed, but an issue through which we can identify the deeper structural flaws of our ways of life, and through which we can become good individuals.

As I have argued in Chapter 1, the common governmental tactic of handing over responsibility to science, economics or a combination of the two will indeed not provoke the social shift and collective actions necessary to address climate change. But, as I will explain, the 'ethical-individual approach' running alongside this 'techno-economic' one, is also flawed. Although an ethical analysis of climate change reveals important concepts and trends, and forces us to examine previously unexamined values and behaviour, the problem with understanding climate change as solely or primarily a matter of ethics is that it advocates and reproduces a globalisation and individualisation of responsibility. This dual trajectory makes responsibility both universal and personal. It disregards political communities and identities, and leaves it up to individuals to decide when and how to act.

I will show that this portrait of the human individual actually replicates – from another angle, as it were – the one painted by the techno-economic school. In this picture, each individual is again presupposed to have access to one 'rational' perspective through which the problem can be assessed. But if we understand humans to hold many diverse ethical perspectives, then to expect these perspectives to translate and harmonise into one unified movement is surely not feasible. This approach mistakenly presupposes, first, that the actions of countless individuals will be coordinated into some sort of harmonious green movement to save the planet. Second, it can result in the depiction of those who don't fulfil their supposedly ethical environmental duties as morally bad. Thus it precludes collective responses to tackling climate change, both by assuming that ethical alignment of people around the world is feasible, and then by excluding those who don't align themselves.

Ethics and Responsibility

Ethics can be defined both as 'an activity of thinking' and as 'a set of values which guide an individual or group in their behaviour'.[1] 'What ought I to do?' and 'How should we live?' are ethical questions, which indicate considerations beyond self-interest. Thus an ethical approach to climate change highlights the values that condition our attitudes towards our environment and our responses to environmental harm, and simultaneously reflects upon these values. The terms 'ethical' and 'moral' are often used interchangeably, but I have chosen to use the former in this chapter. Whereas morality is understood to consist of universal principles and rules, ethics (as I use the term here) is about human values in a particular context. As the context is climate change, I regard the present debate about values as an ethical rather than a moral one.

Various theorists and activists lament the lack of ethical discussion about climate change, for they insist that climate change is essentially an ethical issue.[2] They explain that although the attention of other disciplines is welcome and important for considering the issue, ultimately the judgements about whether and how climate change should be tackled need ethical reflection which cannot be substituted by economic or scientific assessment. The philosopher Stephen Gardiner, for example, points out that just recognising climate change as a problem requires an ethical perspective.[3] He asserts that the issue demands greater engagement of moral philosophy: 'climate change poses some difficult ethical and philosophical problems. Partly as a consequence of this, the public and political debate surrounding climate change is often simplistic, misleading, and awash with conceptual confusion. Moral philosophers should see this as a call to arms.'[4]

Like Gardiner, Dale Jamieson believes that the ethical assumptions and ideas in the background of the climate change debate need bringing to the fore. Jamieson points out that the disagreement between Stern and Nordhaus (discussed in Chapter 1) is not really a quarrel over economic reckoning but is rather a disagreement about values. The two economists discount the future differently in their calculations, which means, in short, that they disagree about how the interests of future generations should be weighed against those of present generations. Jamieson explains that this dispute is clearly a dispute about ethical values.[5] He points out that climate change raises unavoidable ethical questions: 'questions about who we are, our relations to nature, and what we are willing to sacrifice for various possible futures'. Thus the 'management' approach that narrowly restricts itself to economics cannot provide the answers: 'We should confront this as a fundamental challenge to our values and not treat it as if it were simply another technical problem to be managed.'[6]

For Jamieson, by failing to prioritise the environment in the decisions we make, we damage ourselves. Climate change, then, is identified as the symptom of a fundamental flaw in our current human existence, a crisis that demands a radical transformation in our way of life and in our ethical perspectives. The tackling of climate change is not only about saving the environment, since it would also foster human integrity and wholeness: 'Developing a deeper understanding of who we are, as well as how our best conceptions of ourselves can guide change, is the fundamental issue that we face.'[7]

Jamieson claims that the obstacle to proper thought and action about global environmental issues is the dominant set of values, inherited from philosophers who lived in a very different era. Our current value system, he explains, arose in an era of low population density and low technology, and is not suitable for today's changed world. The particular understanding of responsibility found in this system is based on the idea that a harm usually has an obvious and specific cause and therefore assigning responsibility or blame is a straightforward matter. But this model of responsibility, he explains, doesn't work for a complex issue in which cause and effect are remote: 'what we need are new values that reflect the interconnectedness of life on a dense, high-technology planet'.[8] Since climate change is caused by the innocent acts of countless individuals, and the harms will mainly be manifested in the future, it is difficult to blame any persons in particular. This doesn't fit with our normal perspective of an 'urgent moral problem'.[9] Unless we promote a different understanding of responsibility, no one will feel obligated to tackle climate change, and our responses will be wholly inadequate. We need a revision of our everyday understanding of responsibility.

This call for a transformation of our value system and specifically for a new conception of responsibility is echoed by other thinkers. Peter Singer agrees that our current way of thinking needs an overhaul to reflect the fact that the environment can no longer be understood as an unlimited resource, and that responsibility for environmental harm is not easy to assign. As he puts it, 'the twin problems of the ozone hole and of climate change have revealed bizarre new ways of killing people. By driving your car you could be releasing carbon dioxide that is part of a causal chain leading to lethal floods in Bangladesh.... How can we adjust our ethics to take account of this new situation?'[10] And Nigel Dower lists climate change along with various other issues that necessitate a new 'global ethic' which includes trans-boundary obligations: 'The problems of the world, such as absolute poverty, conflict, environmental degradation, climate change, refugees and human rights abuses, require of individuals and states a new sense of global responsibility.'[11]

These theorists draw crucial attention to the challenges and confusions in thinking ethically about global environmental problems. However, as I will go on to show, the new conceptions of environmental ethics they call for are generally characterised by both a globalisation and an individualisation of responsibility. States and political institutions are not burdened by ethical responsibility; in a cosmopolitan move, the state is overlooked and responsibility for climate change is handed to individuals. The burden is place upon the good individual to tackle climate change by carrying out what Michael Maniates calls 'small, individual eco-actions' such as recycling, swapping light bulbs and buying a bike.[12] But this detracts attention from political movements and collective actions. Without political collectives and institutions through which demands for larger, more sustained action can be channelled, individualised conceptions of ethical responsibility can only go so far in actually implementing change. Despite its calls for a radical structural change by individualising responsibility, the ethical response is impotent when it comes to instituting it. What this approach does execute effectively, however, is to discriminate between human individuals as ethically 'good' and 'bad'; as worthy or wrong, green or greedy.

'Thin' Cosmopolitanism

It is often asserted that the globalisation of information and communication, and the growing interconnection of the world, produces a corresponding globalisation of responsibility. This is the perspective of cosmopolitanism, a normative outlook which prioritises the individual over states and communities. As Kwame Anthony Appiah states in his analysis of the term, 'the one thought that cosmopolitans share is that no local loyalty can ever justify forgetting that each human being has responsibilities to every other'.[13] Appiah explains that cosmopolitanism is not an outright rejection of national boundaries, nor the eradication of the special bonds of family and community – but it is a perspective that asserts that all human individuals have basic rights and responsibilities, in whichever part of the world they reside.

As an urgent global problem for which political boundaries and state borders are irrelevant, climate change certainly can appear to call for a cosmopolitan response. Lorraine Elliott explains that environmental harms, including those of a changing climate, are 'displaced' in various ways, so that those who cause the problem are not the ones most impaired by it: 'transactions of environmental harm ... extend the bounds of those with whom we are connected, against whom we might claim rights and to whom we owe obligations within the moral community'.[14] Environmental issues actually broaden the scope of obligation to the entire planet: 'environmental harm deterritorialises (or at least transnationalises) the cosmopolitan community'.[15]

Elliott draws attention to the people who are marginalised ecologically, economically and politically, and demands that they should be acknowledged as moral subjects. She criticises current environmental conventions for concentrating upon relations between countries rather than people; international harm conventions, she explains, should prioritise the consent of individuals not states in the decision-making processes: 'perhaps the most important contribution that cosmopolitan approaches bring to an investigation of the form of environmental harm conventions is the demand for individual and community (rather than state) consent'.[16] However, Elliott doesn't explain by what mechanisms she expects subaltern groups to be brought into the political discussions about environmental issues. Rather than engaging with the political processes of the state, then, she recommends bypassing it, awaiting instead the opening up of the international debates through some unspecified ethical procedure. It is not clear that these debates would not marginalise and coerce individuals to an even greater extent.

Paul Harris also offers a cosmopolitan analysis of climate change, asserting that: "Climate change cries out for a cosmopolitan response. It is a global problem with global causes and consequences.'[17] As Harris confirms, cosmopolitanism is a normative perspective that focuses upon the individual: 'A cosmopolitan approach places rights and obligations at the individual level and discounts the importance of national identities and state boundaries.'[18] To the current state-centric order Harris thus proposes a new 'cosmopolitan corollary' that focuses upon the rights and responsibilities of individuals. While Elliott emphasises the marginalised poor of the developing world, Harris concentrates instead upon the forgotten rich of the developing world, pointing to the growing number of wealthy 'new consumers' in developing countries who have high carbon emissions but are often unnoticed by theorists and policy makers. Harris argues that affluent individuals have a responsibility to tackle climate change, whether they live in the developed or the developing world: 'affluent people *everywhere* should limit, and more often than not cut, their atmospheric pollution, regardless of where they live'.[19]

Harris, to be sure, doesn't recommend the demise or dilution of the role of states, but rather suggests that all states should promote a cosmopolitan ethic and legally obligate their affluent citizens. He expects states to implement this 'cosmopolitan corollary' through various mechanisms including taxation, regulation and education. But what would motivate states to do this? He proposes the creation of global funds raised through taxing greenhouse gas emissions and distributed 'fairly' to those who need help adjusting. But who gets to decide how these funds would be raised and allocated? And who gets to decide who gets to decide? In a paper co-written with Jonathan Symons, Harris considers the distribution of such funds, explaining that adaptation assistance for climate change would be conditional upon the implementation of cosmopolitan principles: 'An adaptation funding institution that is informed by cosmopolitan principles, might … make domestic implementation of cosmopolitan standards a prerequisite for receipt of international assistance.'[20] They note, however, that as yet the international community has not reached any agreement on how to make fair allocations of climate change assistance, and point out that 'achieving agreement on the forms and degree of hardship that warrant adaptation assistance must inevitably be a political process'.[21] Yet is this political process just a straightforward formality? This account downplays the inevitable difficulties in the way of reaching an international political agreement about the securing and distribution of such funds.

The emphasis in Harris's account is upon the ethical obligations of individuals to cut their emissions, especially if the states they live in are not acting to prevent climate change: 'if governments are not fully up to the task ... affluent individuals will have to find it within themselves to act upon cosmopolitan obligations'.[22] So the onus here is on individuals to act independently of political coercion and decision. Both Elliott and Harris importantly draw our attention to groups of people who too often go unnoticed. But both accounts assume that ethical responsibility is adequate to drive a structural change needed to respond to climate change – and on this point, fundamentally, they fail to convince.

The problem of assigning responsibility for climate change is examined by Simon Caney. Caney offers a thorough critique of a principle much cited in understanding the responsibility for environmental harms: 'the polluter pays principle', which allocates the duties to bear the burden of tackling climate change to those who caused the problem by producing the polluting greenhouse gas emissions in the first place. Caney explains that the intergenerational dimensions of climate change make 'polluter pays' an incomplete account of responsibility; high levels of greenhouse gases have been emitted since the Industrial Revolution and therefore the vast number of the 'polluters' are no longer living. Caney rejects the idea that current generations should have to pay the costs of the pollution of earlier generations: 'making current individuals pay for pollution that stems from past generations is indeed making someone other than the polluter pay'.[23] Caney recommends, instead, a set of duties that supplement the 'polluter pays principle' with an 'ability to pay principle'. Here the duties to tackle climate change fall on all, with a special burden on the more advantaged.

Caney, however, notes the problem of how to implement this responsibility, and incorporates into his account a responsibility to build the institutions that encourage acceptance of the burden of paying for pollution; one of the duties he lists is 'a duty to construct institutions that discourage future non-compliance'.[24] But here Caney is simply asserting the responsibility to accept responsibility; nothing is addressed to those who resist it. This responsibility is not based on anything but itself. He is left with this problem, I suggest, because he summarily rejects the 'collectivist interpretation' of the 'polluter pays' principle. Caney claims that the collectivist interpretation, which assigns present generations of a nation-state (or some other sort of collective) the burden of paying the costs generated by earlier generations, is flawed since it has difficulties justifying the political obligations of citizens of a state who have not actually given their consent to the policies of that state. This problem of political obligation is a lasting point of contention in political theory, for it is difficult to see why citizens who had no choice in their membership of a state, should be obligated to it.[25] But to write off the existence of political obligation entirely, as Caney does, leaves him with the difficulty of justifying the collective responsibility to tackle climate change. And if political obligation has been scratched it is not clear how Caney expects to secure adherence to the political institutions he suggests need to be built to ensure compliance.

A responsibility embedded within citizenship would be underpinned by the stronger ties of membership in, and identification with, a political collective. We would be burdened as a collective to tackle problems resulting from greenhouse gas emissions of previous generations, a responsibility that would be met through collective decision and action. The individualist interpretation of environmental responsibility that Caney prefers is underpinned only by personal conscience, leaving it entirely up to the individual to comply – or not.

'Thick' Cosmopolitanism

The cosmopolitan ethical obligations considered so far appear to lack any motivating force connecting these obligations to actions. As Andrew Dobson explains, these cosmopolitan accounts base our ethical obligation to act against climate change upon nothing but our common humanity, but while these 'thin ties of humanity' might persuade us to think like cosmopolitans, they are not enough to get us to *be* cosmopolitans.

Dobson notes the difficulty of the 'tyranny of distance' that leads individuals to be less concerned about others who are 'distant' than they are about those who are close to them. Humans prioritise their friends, family and neighbours over those on the other side of the world. Dobson explains that '"nearness" ... has a bearing on our motivation to respond to the prompts of obligation. What cosmopolitanism requires is a "nearness" to the vulnerable, suffering, disadvantaged others, and the recognition that we are all members of a common humanity seems not to bring such others near enough.'[26]

What Dobson calls 'thin cosmopolitanism' demands that we act as Samaritans, that we help others from no other motivation than our moral obligations deriving from shared humanity. But the 'universalism of Samaritanism', he explains, is too weak; it is good to act as a Samaritan, but it is not wrong not to. Dobson makes the following claim: 'The reason why we feel especially moved by the act of the Good Samaritan in assisting the poor unfortunate by the side of the road is that the Samaritan was not at all responsible for his injuries.'[27] He goes on to claim that if we see someone as causally responsible for a problem the obligation to act is 'thickened': 'If, on the other hand, the Samaritan had been implicated in the man's suffering in some way or another, we would expect him to go to his aid.'[28]

Dobson goes on to advocate a different interpretation of cosmo-politanism – a 'thick-cosmopolitanism' or 'post-cosmopolitanism' based upon causal responsibility that makes us nearer to distant others. His theory is still nonetheless a version of cosmopolitanism, in that responsibilities arise not from the vertical connections between individuals and the state, but rather from horizontal links to other individuals, relations that disregard nation-state boundaries. But rather than the thin links of morality, Dobson says that these links have a material thickness which exists through relations of cause and effect: 'We are more likely to feel obliged to assist others in their plight if we are responsible for their situation – if there is some identifiable causal relationship between what we do, or what we have done, and how they are.'[29] He understands this material thickness to be displayed by 'ecological footprints'. Everyone in the world has an ecological footprint: for some it is small; for most in the developed world it is large and unsustainable. The ecological footprint reveals material and asymmetrical relations that underpin our ecological responsibilities.

This material causal responsibility then 'thickens' the responsibility to act to tackle climate change and is the basis of Dobson's conception of 'ecological citizenship'. Dobson suggests that this makes the provision of adequate help to those in need due to the problems of climate change more than a matter of providing aid, but rather a matter of justice. 'Causal responsibility produces a thicker connection between people than appeals to membership of common humanity, and it also takes us more obviously out of the territory of beneficence and into the realm of justice.'[30]

However, although Dobson has perhaps 'thickened' the obligation, he has not made it any more imperative – as regards acting upon it – than the 'thin cosmopolitanism' he criticises. There is still no necessary link between a thick obligation and the fulfilment of that obligation. Dobson seems to take for granted the predisposition of individuals to act upon their environmental responsibilities, but he offers no mechanisms binding individuals to these responsibilities. He may have shifted responsibility away from the 'thin' bonds of humanity to the 'thick' obligations of justice – but, still, without any political obligations to ensure that justice is done, it is not clear why individuals should act justly and carry out their moral obligations (thin or thick) to reduce their ecological footprint.

Dobson seems to presuppose that once individuals are made aware of their ecological footprint, they will accept their thick responsibilities. But this presupposes that they see the world, and responsibility, in the same way as he does. Dobson begins from a particular conception of responsibility – he assumes that it is thickened by causality. But not everyone would concur. For example, the Good Samaritan might not be causally responsible for the injuries of the unfortunate, but he or she might well regard themselves as nevertheless burdened by a strong responsibility to help. Why does a causal link thicken it? Dobson assumes, already, that responsibility is in some way linked to causal responsibility. Many may agree, but this would involve their regarding the world in the same way, and there is no reason to believe that everyone does. This conception of responsibility is no thicker, in the end, than thin cosmopolitanism. Both rely on individuals sharing the same notions of responsibility.

Even if individuals do act to reduce their ecological footprint, presumably there are many different ways to do this. The precise implication of the responsibility to ensure that one's ecological footprint is sustainable, as Dobson explains, is not already decided: 'I do not propose to outline a manifesto for "green living" .... The obligation is evidently radically indeterminate.'[31] The specificity of the action is left up to the individual to decide. Dobson assumes that, once the thick obligation arising from a causal responsibility has been revealed, everyone will act in harmony to reduce their ecological footprint. Yet it cannot be assumed that there is one obviously 'correct' version of living sustainably and tackling climate change. And if individuals put different versions into practice, then it seems possible that the many diverse individual actions might well cancel each other out, or hinder each other, or at the very least not work as well as if they were coordinated from within an overarching polity. For example, while some might pursue nuclear power as an environmentally friendly source of energy, others might oppose such policies for the very reason that they threaten the environment. Isn't a collective decision and action needed to ensure that individuals coordinate their actions?

Cosmopolitans marginalise the importance of any collective decisions and any political debate in reaching these decisions. Presumably, then, they believe that no explicit agreement upon the action that should be taken to 'live sustainably' is needed. They don't necessarily rule out collective action, but they do leave it up to individuals to decide whether to participate in it. They seem to assume that there is some sort of 'invisible hand' at work coordinating the actions and lifestyles of individuals into a harmonious green movement to combat climate change. This assumption rests upon a presupposed human capacity that can acknowledge a responsibility to reduce their ecological footprints, intuitively calculate the right coordinated actions to take to do so, and prompt the performance of these actions. It is far from certain, however, that any such human capacity exists; I suggest that cosmopolitans underestimate the role of political community in supporting and ensuring these individual actions.

Dobson does suggest that a political community of 'ecological citizens' arises through the material relations of the ecological footprint, a community which is constituted solely by 'a space in which political obligation operates'.[32] But no common institutions, language or identification unite this community; ecological citizenship is, as he puts it, a 'citizenship of strangers'.[33] Surely a political community needs more from which to produce and sustain a shared understanding of justice and responsibility? Tim Hayward observes that a political community with shared principles of justice involves not just material but communicative relations: 'purely material relations are not social relations and cannot on their own generate any principles of justice'.[34]

Cosmopolitanism – thin or thick – hands the burden to individuals, which is inadequate but also unfair. We cannot rely upon individuals – as individuals – to act decisively and harmoniously to tackle climate change, nor should we expect them to do so. We need to shift responsibility back to political collectives and institutions. I argue that the action and policies required to reduce the size of ecological footprints and to tackle climate change require the decisive commitment of a political collective. Cosmopolitans, however, have led the way in inhibiting any focus upon states. And, as we will see in the next section, governments have responded by willingly shunting the burden of climate change to individuals.

Green Consumers

Plastic carrier bags, according to the Department for Environment, Food and Rural Affairs (DEFRA), are a symbol of our 'throwaway society' and their manufacture produces unnecessary carbon emissions. But rather than introducing laws to solve the problem, the burden for lowering their use is given to the individual consumer. The DEFRA website provides tips for remembering reusable plastic bags (keep them in your handbag!) and a statement reads: 'We expect retailers to take responsibility and cut down on the number of single-use carrier bags they hand out, but the ability to take action also lies with consumers, who can decline to accept them in favour of reusable alternatives.' It goes on: 'If results do not improve we will consider additional measures, including legislation.'[35] The starting point, then, is the individual consumer, not the legislation. This focus upon the individual is reproduced in the introduction of what are sometimes referred to as 'soft' policies to tackle climate change, such as 'save energy' campaigns and recycling schemes.[36] The individual is regarded here as a consumer; not the economic consumer considered in the last chapter but the ethical consumer who buys 'green'. We are urged as individuals to check our personal carbon footprint and to install 'microgeneration' ('the generation of low, zero or renewable energy at a "micro" scale').[37] Tackling climate change is regarded as the perfect role for the 'Big Society', which rolls back the state and leaves it up to individuals to engender change through voluntary work. Greg Barker, minister of state for energy and climate change, sees everything falling into place while profits benignly rise for all concerned: 'Community energy is a perfect expression of the transformative power of the Big Society. With the right combination of incentives and freedoms, community groups, businesses and organisations can get together to build a cleaner, greener future. They can generate their own heat and electricity, and their own profits, and as a by-product, help the UK to save energy and help to cut carbon emissions.'[38]

Politicians assert that climate change is a moral issue, rather than a political one, and fail even to attempt the instigation of collective political protest. On the back cover of the DVD for Al Gore's film *An Inconvenient Truth* is a list of '10 things to do to help stop global warming'. They include 'plant a tree' and 'drive less'.[39] Chris Cuomo mourns the lack of *political* actions on this list: 'Every item on the list', she notices, 'is an individual action to be carried out in the personal sphere, and there is no mention of more political options such as "pressure your senator" or "rally against mountaintop-removal coal mining"'.[40] This is problematic because, as Cuomo explains, the carbon reductions that individuals are able to achieve are limited. First, household consumption and personal transportation produce only 20 per cent of the total greenhouse gas emissions.[41] Second, most people have little control over energy source options: 'popular environmentalist discourse tends to emphasise personal responsibility, or the need to shift desires on the demand side, but instant replacements for existing technologies, materials, and forms of transportation are not readily available everywhere'.[42] Cuomo argues therefore that an adequate response to climate change demands the acceptance of responsibility by governments and large corporations. But the tendency she notes is part of a more worrying trend towards the ongoing shifting of responsibility to the individual.

As Michael Maniates points out, 'contemporary environmental action has tilted toward an unpromising politics of guilt focused on the individual behaviour of the many, rather than engaging politics of structural transformation that mobilizes the most committed'.[43] Maniates sees this individualisation of responsibility as an increasingly dominant response to climate change: 'This response half-consciously understands environmental degradation as the product of *individual* shortcomings ... best countered by action that is staunchly *individual* and typically *consumer-based* (buy a tree and plant it!) It embraces the notion that knotty issues of consumption, consumerism, power and responsibility can be resolved neatly and cleanly through enlightened, uncoordinated consumer choice'.[44] But, as he points out, this is not an adequate response to environmental issues; rather, it is a flight from politics: 'at a time when our capacity to imagine an array of ways to build a just and ecologically resilient future must expand, it is in fact narrowing'.[45] Individualisation precludes any opportunity for the empowering political experiences of acting politically and collectively. We become wrapped up in our own ethical choices and insulated from the broader political issues.

For Maniates, this individualisation is a result of various factors, including the dominance of neo-liberalism and the avoidance by environmental groups of political confrontation and the complexity of the issues themselves. But as the individual is increasingly atomised, this process feeds upon itself. What is needed is a collective political response that ensures institutional change. As Maniates puts it, 'confronting the consumption problem demands ... the sort of institutional thinking that the individualisation of responsibility patently undermines. It calls too for individuals to understand themselves as citizens in a participatory democracy first, working together to change broader policy and larger social institutions.'[46]

Maniates explains that the dominant response is to obsess over consumer choices. Individuals are encouraged to think of themselves as ethical consumers. Here we are, then, back at the depiction of individuals as consumers first and citizens second. The ethical approach contains a similar picture of the individual to that of the techno-economic approach. In both approaches, people are regarded as rational consumers, but rather than a self-interested calculator, the individual is now environment-interested. Again there is a 'right' way to behave and to consume; to be 'good' one should buy green products. Instead of being categorised as 'rational' or 'irrational', here people are divided between 'good' or 'bad'.

Detonating (In)Difference

The individualisation of responsibility is echoed in campaigns that bypass states and international organisations entirely and directly target populations demanding individual ethical commitments. The 10:10 campaign, for example, encouraged the personal pledges of individuals to reduce their carbon footprint by 10 per cent in the year 2010. This campaign was launched by the *Guardian* just after the COP UN Climate Change summit in Copenhagen. 'The politicians failed in Copenhagen,' the headline read, 'so now it's up to you.'[47] The campaign has been globalised, reaching 171 countries.[48] The focus here is lifted off governments and political power, and placed instead upon individuals and their moral consciences. The claim is made that 'while the politicians bicker and backbite thousands of people, organisations and companies have been quietly taking the battle against climate change into their own hands'.[49] Critics might ask for a way of holding individuals and companies to their pledges, but this would be to miss the point. It was not supposed to be an enforceable policy; it is an individual ethical commitment, and the organisers are aware of this fact. The aim of these sorts of schemes is to cultivate an ethical commitment that is as binding as an economic incentive or the rule of law. The important question is whether the ethical approach advertised by such a scheme as 10:10 helps or hinders any move to combat climate change. By prioritising the individual it can detract attention from the political efforts to engage the state in tackling climate change. As Andrew Dobson declares, '10:10 is about changing light bulbs rather than changing society'.[50]

The 10:10 campaign featured the release of a short film intended to encourage people into taking personal action to reduce their carbon footprint. Rather than a sombre or dry documentary, it used black humour to make its point. In its graphic scenes, detonators are used to explode those who are indifferent to the problem of climate change (including two school children) into bloody pieces. It communicates its message clearly, and its title *No Pressure* is an ironic comment on the film's highlighting of the burden on individuals to act.

After the immediate outcry of protest, the film was withdrawn, with apologies by everyone involved and pleas that they had 'seriously misread' the mood. But written by Richard Curtis, directed by Dougal Wilson, featuring the film star Gillian Anderson and England footballer Peter Crouch, with music donated by Radiohead, it was no minor or amateur undertaking. I suggest that those behind the film didn't go wrong at all, but rather took the campaign's portrayal of the issue to its extreme. The film was part of the ethical-individual account of climate change, an account that portrays individuals as good or bad, as worthy of moral praise or condemnation. This account regards climate change as an ethical responsibility for the individual to tackle; those that flout this responsibility should be removed from the social sphere, extracted with a similar ease to the pressing of a red button on a detonator.

While it might indeed be perceptive and helpful to highlight the important ethical questions and obligations that arise with an issue like climate change, the presumption that ethics is not just one important part of the response to climate change but rather should dominate the discussion is seriously problematic. For Stephen Gardiner, 'ethical questions are fundamental to the main policy decisions that must be made, such as where to set a global ceiling for greenhouse gas emissions, and how to distribute the emissions allowed by such a ceiling'.[51] As I will go on to argue, however, these decisions are more usefully understood as political decisions.

Gardiner suggests that the complexity of the problem of climate change makes it perfectly convenient not to do anything to tackle it. He regards the issue as placed in the eye of a 'perfect moral storm'. This storm makes us vulnerable to moral corruption and imperils our ability to act in an ethical way: 'climate change may turn out to be a perfect moral storm ... its complexity may turn out to be perfectly convenient for us, the current generation, and indeed each successor generation as it comes to occupy our position ... it provides each generation with the cover under which it can seem to be taking the issue seriously'.[52] But this, I believe, is precisely why we need political and collective decisions.

The ethical approach, I contend, relies upon the individual making the 'right' ethical choices. This echoes the techno-economic approach, but where the ethical approach differs is by conceptualising the 'right' behaviour not just as rational but as good. The individual is given the power to make an ethical adjustment of society, to avert the tragedy that otherwise awaits the planet. But the depiction of climate change as solely an ethical issue helps only the assuaging of our own consciences, and our self-persuasion that we are 'good' because we are 'green'. The problematic corollary to this is that those who for whatever reason do not adhere to green behaviour and values are immediately categorised as 'bad', and are automatically excluded from ethical consideration and political debates about the issue. As I will explain further in Chapter 5, this marginalisation through ethical vilification precludes the possibility of revitalising politics through disagreement. Ethical theorists importantly draw attention to the way in which climate change reveals deep and problematic tendencies within society. But what is also of interest is this very depiction itself; what does the account of climate change as an ethical problem reveal about our current socio-political world? Does it indicate an increasing individualisation and moralisation of issues that could be seen alternatively in more political and collective terms? In the next chapter, then, I look at the salutary shift in focus from individual to collective found in green republicanism.

3 Responsible Citizens
The Green Republican Approach

From Pinprick to Daub

Viewed from the perspective of the planet's atmosphere, the human individual appears as an infinitesimal pinprick, one that flickers only briefly with life before being subsumed back into oblivion. Climate change, as a problem understood at this level, has a tendency to make us feel irrevocably irrelevant and to make our actions seem ultimately futile. For what is the action of one individual against a problem created by a species of seven billion members, all apparently hell-bent on gleaning their share of planetary resources from under its crust and out of its oceans and forests? Climate change, it seems, cannot be micromanaged. It needs action on a wider, longer and more significant scale. Many faithful and fervent green activists end up disillusioned by 'green wash' and disheartened by gloomy prognoses. As one such previous 'eco-worrier' now suffering 'green fatigue' states: 'Only if we look to collective action, and the brute force of government legislation, will we be able to make a meaningful difference.'[1] To respond to climate change we need to collate the tiny insignificant pinpricks into one giant green daub.

Although very different in their depiction of the problems of climate change and its potential solutions, both the approaches examined so far have been found lacking. Both the 'techno-economic' and 'ethical-individual' approaches forget the role of the political collective. Both expect the individual to tackle the problem of climate change alone, unbidden by others and oblivious to power relations. The techno-economic approach reduces the human individual to a self-interested calculator, only motivated by economic concerns, and guided by the invisible hand of the market into making the adjustments necessary to combat greenhouse gas emissions. The ethical-individual approach, in contrast, elevates the human individual to a rationally thinking 'good green', driven by conscience to tackle the warming climate like a frontiersman up ahead of a deeply flawed civilisation. Although these are very different depictions of human beings and the world they live in, I argue that the social atomisation inherent in both approaches marginalises the possibility of collective political action that is crucial in responding to climate change. A promising alternative is examined in this chapter. Here I consider 'green republican' accounts, which regard the individual primarily as a community-oriented citizen.

In these accounts, attention is shifted from individuals to communities. Not the individual, but the community is now expected to instigate the move towards a climate-friendly society. Green republicans assert the existence of a commonly held understanding of 'the good' within which they embed a substantive 'green' component. Rather than self-interested consumers, individuals are expected to be responsible and virtuous citizens who actively participate in the creation and enactment of this green common good which is, importantly, understood to involve environmental sustainability.

The problem with green republicanism, however, is that it presupposes the existence of one overarching conception of 'the good'. Who decides the substantive content of the 'common good' and the citizenship virtues and responsibilities required to uphold it? Does this approach result in an authoritarian demand for compliance with a particular set of values and behaviour? What of the gender and class identifications that may result in varied understandings of and relationships to climate change? Although green republicans appear to be sensitive to the diversity of society, ultimately, I contend, they assume the possibility of a way of life with which everyone agrees, and their accounts rest upon a virtuous citizen who exhibits certain dispositions, including the rationality to agree upon the substantive content of the common good. The agreement on the common good demanded by green republicanism is antithetical to pluralism; it actually hinders democratic decision making and, ultimately, action on climate change.

The Turn to Citizenship

Many environmentalists are all too aware of the problems of the individualistic approaches considered in the previous chapters, and recognise that a climate-friendly society demands the engagement of people who think beyond self-interest and are motivated by concern for the common good. After all, many people do, at least sometimes, act in a way that prioritises the good of the community over their own personal desires. It is this more *citizenly* attitude, it is argued, that is required in combatting climate change. As Dobson and Bell point out in their book on environmental citizenship, 'it is surely a fantasy to think that sustainability can always be a win-win policy objective, in which each gain for the common good will also be a gain for each and every individual member of society'.[2] Such a turn to citizenship is expressed in the manifesto of the Green Party: 'As citizens we think of the good of everyone and of the future, and not just what we think is good for ourselves, now. Creating a fair and sustainable society is a job for government at all levels – but it is also a job for us as citizens ....' The Green Party distances itself from the atomistic presuppositions of mainstream political parties, emphasising the role of the government and the community and the importance of quickening a sense of the public good: 'Fairness and sustainability require a public effort. Millions of uncoordinated actions aimed at maximising individuals' private interest will not bring about the society we want and deserve. Only citizens, aided by a Green Government, can do that.'[3]

As a citizen, a person is expected to move beyond selfish concerns and a narrow perspective on the world. Citizenship, as Anne Phillips explains 'is a concept that deliberately abstracts from those things that are particular and specific and seems to shift us onto a higher terrain ... citizenship often propels us towards an ideal of transcendence, a greater collectivity in which we get beyond our local identities and concerns'.[4] Hence there has been a 'turn to citizenship' in much of the environmental politics literature.[5] This turn has produced a body of work on green citizenship that injects environmental sustainability into the concerns of the community-oriented citizen.⁶ There are various versions of green citizenship but all emphasise the existence of citizenship *responsibilities* that should not be overshadowed, as often seems the case today, by the package of *rights* that comes with the status of citizenship. As a green citizen I become aware of my responsibilities towards the social and natural world, and am not merely focused upon my rights to it: 'the environmental citizen's behaviour will be influenced by an attitude that is – in part, at least – informed by the knowledge that what is good for me as an individual is not necessarily good for me as a member of a social collectivity'.⁷

The dominant conception of citizenship is the liberal version expressed in T. H. Marshall's influential mid-twentieth-century account, in which citizenship is regarded as an equal status conferred on all members of a population; citizenship is thus supposed to overcome the various existing inequalities (such as class) within any society.⁸ Marshall suggests that there are three elements to citizenship – civil, political and social – and that each of these elements grew through the establishment of a corresponding set of citizenship *rights*. Civil, political and social rights aim at securing liberty, access to political power and a basic standard of living for all individuals. How citizens use these rights and what they choose to do with their lives, however, is very much their own personal choice. Liberals prioritise the freedom of the individual from interference by others and by the state; people have very different ideas of what is good in life and they should be given the freedom to follow their particular idea. The individual is regarded as sovereign and the community does not have a stronger collective claim. For liberals, therefore, any substantive conception of 'the good' is regarded as privileging one particular viewpoint over another and curtailing the individual's freedom. The only thing that liberal citizens are expected to agree upon are political procedures and a conception of justice, not on anything more substantive than that; 'the right' has priority over 'the good': 'conceptions of the good are radically divergent, and there is no prospect of people coming to agree about what is of ultimate value to them. In their capacity as citizens, by contrast, people are capable of reaching agreement on principles of justice that will govern their political arrangements.'[9] Liberal accounts of citizenship, in short, focus upon citizenship *rights* and there is less emphasis on corresponding *obligations*.

This liberal model of citizenship is sometimes called 'private' or 'passive' citizenship and it is criticised by many for prioritising rights and neglecting the important component of citizenship responsibilities.[10] In their survey of citizenship theory Will Kymlicka and Wayne Norman write: 'There is increasing support ... from all points of the political spectrum, for the view that citizenship must play an independent normative role in any plausible political theory and that promotion of responsible citizenship is an urgent aim of public policy.'[11] As Kymlicka and Norman indicate, the emphasis upon citizenship responsibility as an important component of citizenship has come from various quarters, and one such is the rising awareness of environmental issues such as climate change. Environmentalists are one key group of thinkers who are trying to rebalance the asymmetries of liberal citizenship. Mark Smith and Piya Pangsapa explain how theorists of green citizenship make an important contribution to the debate: 'For over two centuries, citizenship has been fixated upon rights and entitlements, glossing over duties and obligations ... ecological citizenship is part of a new generation of kinds of citizenship that take the politics of obligation seriously.'[12] The important point made by theorists of green citizenship is not simply that citizens have environmental responsibilities, but that they have numerous responsibilities that demand a more active contribution to society than liberal models would suggest. Issues such as climate change, it is hoped, may help transform dominant conceptions and perceptions of citizenship for the better, and not just in terms of the environment.

An important and enduring alternative to the liberal model of citizenship is offered by republicanism, which centres upon the notion of the citizen as a responsible and active participant in the political life of the community. Republicanism upturns the prioritising of rights over responsibilities and promotes 'the common good' as something that unifies the members of the community; something that every member draws from and in turn contributes towards: 'A citizen identifies with the political community to which he or she belongs, and is committed to promoting its common good through active participation in its political life.'[13] Many environmentalists have embraced the republican model, redrawing its notions in a bold shade of green. The next sections probe the detail of this redrawn conception.

Green Republicanism

Republicans often refer to Aristotle, for whom man is only able to be truly fulfilled by actively participating in the public realm. 'Man is naturally a political animal,'[14] he claims, asserting that man's highest calling is as a citizen; only by contributing to the political life of the community can he reach *eudaemonia* or happiness. The well-being of the individual is subsumed within and secondary to the well-being of the community or city: 'the city, which must be one, and this every citizen must have a share in'.[15] Citizenship in Ancient Athens involved an active role in the running of the city-state, by taking part in government and holding public office.

As David Held notes, today's model of citizenship, in which political participation is commonly limited to occasionally voting for representatives to make decisions for us, would be incomprehensible to Aristotle: 'the ancient Greeks would have found it hard to locate citizens in modern democracies'.[16] Held shows that although republicanism and its ideal of the active citizen has not been a typical regime through human history, periodically it has been revived, by the city-states of Italy in the late eleventh century and, in different ways, by the philosophers Machiavelli and Rousseau. There are different strands of republicanism but, in general, republicanism demands a richer, more active role of the citizen, constituting a call against the political passivity often lamented today. As Kymlicka and Norman point out, however, republican views surely go against the outlook of most, for whom 'political participation is seen as an occasional, and often burdensome, activity needed to ensure that government respects and supports their freedom to pursue ... personal occupations and attachments'.[17] They argue that this is not because public life has become impoverished, but rather because private life has become so much richer. Yet the ostensible enrichment must be due at least in part – as Kymlicka and Norman readily acknowledge – to rising consumption and materialistic lifestyles, the very trends that are challenged by many greens and people concerned by climate change. This common opposition to the elevation of the private consumer over the public citizen flags up the compatibility of republicanism and green thinking.

Stuart White asserts that many of the ideas expressed by republicanism do indeed dovetail with both environmental politics and leftist principles, and he suggests therefore that republicanism is revived and reimagined in a new 'red-green' form: 'By engaging with republican thought, red-green politics can perhaps come to a better understanding of its own underlying values and principles. This, in turn, can provide new insights for red-green politics.'[18] White explains that the 'passive' citizenship encouraged by contemporary democratic forms results in a withdrawing from politics and the public realm, and an individualisation that undermines the active freedom advocated by republicans. In such a passive, individualised society it is 'hard to see how such a society could ever summon up the collective will to confront a challenge like climate change'.[19] A just and sustainable society, White believes, demands the civic virtue that motivates people to consider others and participate in the common good.

John Barry agrees that the ideas and ideals of civic republicanism overlap neatly with those of green politics and argues that since its vocabulary is widely recognised and understood, civic republicanism could be used as a language through which to articulate green principles and to challenge current unsustainable development paths. As he points out, despite the increasing public awareness of ecological problems and the growth of a global environmental movement over the last 30 years, there has been no significant greening of society: 'the gap between awareness, discussion and knowledge of ecological damage and ameliorative action by politicians, governments of individuals grows'.[20] Perhaps what is needed, then, in today's climate-threatened world is a greater emphasis on the environmental responsibilities of citizens and the virtues needed to nurture them.

Barry is clear that while being green doesn't necessarily demand the total eclipse of the consumer by the citizen or the complete quitting of materialistic lifestyles, it does demand the reassessment and adjustment of such lifestyles. This transformation demands sacrifice that can only come about through an awareness of the community and its dependence on a sustainable and sustaining natural world. Barry rejects the conventional reduction of citizens to voters and taxpayers, depicting them instead as active participants in moving to a greener, more sustainable society. Green republican citizenship is not a formal status that is given but a practice that must be learned and continually exercised, requiring rather more than occasional visits to the polling station. Green republican citizens are not self-interested consumers who demand and receive certain rights, but virtuous members of the community who actively contribute to 'the common good'. For Barry, republican virtues shift the focus from 'having' to 'doing': 'the good of which green virtues partake is one in which human well-being is understood as constituted by action rather than possession or consumption'.[21] Through the active participation in politics demanded by the republican model, Barry believes that people will be transformed into 'ecological stewards'. He suggests that although today most individuals don't have direct contact with nature – through agricultural work for example – people can nevertheless become more aware of their natural environment through taking part in the political processes of managing it, and will become in this way more environmentally conscious: 'modern ecological stewardship is premised on the assumption that sufficient knowledge of the world can emerge from the experience of being involved in the political process of collective ecological management'.[22]

Barry claims, then, that republicanism tempers the arrogant anthropocentricism that is common today, which often blinds people to environmental harms. For republicans, the individual is ever vulnerable and dependent upon a particular historical community. Their aim is therefore to build socio-political practices and institutions that secure equality and freedom for individuals, and demand that these are supported by the conscious and collective participation of human beings: 'The republican project is to create a secure home for free men and women ... and this will not occur "naturally" but only by active citizen political action.'[23] The particular historical community is also, Barry notes, always embedded within a particular natural environment, and therefore he enlarges the scope of our vision to include the natural world too: 'civic republicanism acknowledges that environmental conditions shape both the character of the citizens and the possibilities for political action open to them'.[24]

Patrick Curry also agrees that republicanism can be usefully interpreted from a green standpoint. The 'ecological republicanism' he proposes actually expands membership to the natural world for a richer conception of community; he suggests that non-human nature can be included as part of an ecological community, as it is a constituent of its structure and has an awareness of that community. Both humans and non-humans work together – consciously and non-consciously – for the good of that community. He suggests that an 'ecological *virtù*' helps the community survive and flourish: 'In so far as the common good of any human community is utterly dependent – not only ultimately but in many ways immediately – upon ecosystemic integrity (both biotic and abiotic), that integrity must surely assume pride of place in its definition."[25] Curry believes that the environment should be put at the heart of public policy and that, for this to happen, the non-human world must be given greater consideration: 'it is vital that the common human good must give way to that of all life – including, but no longer restricted to, human life'.[26]

Whoever is included in the community and whatever political participation is understood to mean, green republicanism corrects the problematic individualisation of the approaches examined in the previous chapters. It emphasises the embeddedness of the individual in a community and in an environment, and it acknowledges the key role of the collective in a move to a sustainable world. Rather than expecting individuals, as consumers or as moral agents, to take the burden of tackling climate change, green republicanism suggests that environmental sustainability is a matter of the common good that all citizens contribute to and gain from. By promoting a climate-friendly world as in all our interests, and by encouraging active participation in moving towards it, this approach certainly appears to have much greater potential to tackle climate change – as well as to revitalise political life through the engagement and participation of citizens.

However, there are difficulties with green republicanism that I believe hinder its usefulness in approaching the complex issues of climate change. Notions of 'the common good' and eco-virtues rest upon and reiterate some problematic presuppositions.

The Common Good?

Republicanism, as we have seen, centres upon a notion of 'the common good' that individuals contribute to and gain from. But in a diverse society this notion of a common good is extremely problematic. For many, the very notion of any overarching 'common good' makes republicanism guilty of the unforgivable error of overriding individual liberty. Although she welcomes the richer account of citizenship advanced by civic republicanism, Chantal Mouffe, for example, is concerned about the implication of a common public good for pluralism. She explains that the valuable contribution that liberal ideas make to our societies today is to allow individuals the freedom from one overarching idea of the good life: 'a modern democratic political community cannot be organised around a single substantive idea of the common good. The recovery of a strong participatory idea of citizenship should not be made at the cost of sacrificing individual liberty.'[27] As she puts it: 'Modern democracy is precisely characterised by the absence of a substantive common good..'[28]

For green republicans, this problem raises some seemingly difficult questions. How can they justify the imposition of *one* version of a sustainable way of life across the myriad of different cultures and creeds, and all the ethnic and ethical difference of a plural society? Who is to say that a 'green' world is a better world? Should climate change be addressed before world poverty, for example? Some argue that these two goals are not mutually exclusive, but surely one goal takes priority. Even if it might be argued, as indeed I agree should be, that a sustainable world is in everyone's best interest, how might it be decided what a 'green' common good looks like? What, anyway, does sustainability mean? There are differences in conceptions and prioritisations of climate change not only between nation-states,[29] but within them.

Recent versions of republicanism, however, seem able to sidestep this criticism, actually locating the freedom of the individual within active responsible citizenship. Republicans have a very different understanding of freedom to liberals. While liberals understand freedom as non-interference (often called a 'negative freedom') republicans understand freedom as non-domination. As Philip Pettit explains, republicans believe that freedom entails the absence of another agent's capacity to interfere with them, a capacity that is oppressive even if it is never actualised: 'with freedom as non-domination, a person loses freedom to the extent that they live under the thumb of another, even if that thumb is never used against them'.[30] Freedom as non-domination demands that individuals do not have to defer to others, and are ensured 'the ability to look others in the eye'.[31]

Republican freedom is thus a more positive conception of freedom that requires individuals to participate as active citizens in a self-governing community. For Quentin Skinner, it is important to realise that people cannot be free in a community that is externally conquered or internally corrupt, and the only way to ensure that a community is free itself is through the continual vigilance of its members: 'the maintenance of a free way of life requires continual supervision of, and participation in, the political processes by the whole body of citizens'.[32] Republicanism, formulated in this way, suggests that freedom isn't compromised by the common good, but rather is actually entwined with it. Skinner points out that this looks very much like a paradox: 'We can only hope to enjoy a maximum of our own individual liberty if we do not place that value above the pursuit of the common good … the sole route to individual liberty is by way of public service.'[33] What counts here, then, is participation; republicanism of this ilk doesn't assert the existence of a pre-determined common good or shared purpose, but rather an acknowledgement of citizen responsibility, a responsibility that involves not adhering to a pre-defined common good but rather contributing to its substantive meaning through political participation; the common good is given content through a discussion of engaged citizens with various different understandings of the good: 'The republican conception of citizenship', for David Miller, 'places no limits on what sort of demand may be put forward in the political forum.'[34]

This argument allows Barry to claim that green republicanism does not promote one fixed interpretation of 'the common good'. Indeed, he is adamant about it. The version of green republicanism he promotes 'seems more appropriate for a modern pluralist political context in that it is not wedded to the promotion of one conception of "the good life"'.[35] He argues that green republicanism actually keeps the substantive content of the common good open for discussion. Barry acknowledges the lack of any 'truth' about the majority of environmental problems; he argues that rather than solutions from experts what is needed are inclusive discussions through which 'coping mechanisms' can be created: 'general uncertainty and disagreement about the causes, extent and possible remedy for social-environmental problems underwrite the necessity for democratic, open-ended decision-making procedures'.[36] However, although Barry believes that the inclusive political discussion that republican citizens participate in should not be seen as aiming to discover the 'right' answer, he suggests that it can nevertheless attempt to create agreement on the common good: 'green politics ... is not geared towards the discovery of some scientific or metaphysical *truth* regarding social-environmental relations, but is rather concerned with the creation of agreement in respect to those relations'.[37] In his latest book John Barry relinquishes the aim of sustainability for the more realistic goal of reducing 'actually existing unsustainability' since 'there is more chance of agreeing on what is unsustainable than what is sustainable'. [38] Barry, then, believes that the content of the common good can eventually be decided through political discussions involving a plurality of perspectives.

Barry assumes that a diverse citizen body will, inexorably, arrive at the agreement that the common good is a sustainable society (or not an unsustainable one). He argues that through the political participation demanded by republicanism, individuals become aware of their environment and in this way are transformed into environmental stewards, who are concerned about their natural world and wish to sustain its existence for the common good. But can we assume that an awareness and appreciation of the environment will result in a similar perspective to others about what sustainability *means*? Barry and Curry are right to acknowledge the embeddedness of human communities within their social and natural environments, but I believe that this will heighten an inevitable plurality of viewpoints. While establishing an awareness of others and the natural world may make many realise that the environment is, itself, a common good, this does not mean that there is any agreement on how to sustain it. There is not one Nature, but many natures.[39] So, if there is no truth or correct solution to an issue such as climate change, then why should we expect individuals to arrive at consensus on the common good? I suggest that, despite their protests to the contrary, green republican accounts *do* rely on a more substantive common good, although this substantification, it is true, does not occur prior to the discussion. The precise content of the common good is left open by green republicans, but they nevertheless assume that it will be closed by the agreement of virtuous and responsible citizens. It is simply expected that citizens will reach an agreement that a sustainable way of life is in the common good and, further, they will reach agreement on what this way of life looks like. No consideration is given to the possibility that there may be those who are left out of the discussion and that there may be those who don't agree.

Whereas in his account of republicanism Skinner emphasises the importance of citizen responsibility in upholding individual rights and empties the common good out of any substantive content, green republicanism fills it in with an agreement about sustainability. For Skinner, the common good demands the absence of any imposition of its meaning; for White, Barry and Curry, the common good is a climate-friendly world (or, in Barry's case, a not unsustainable one). They slip from what we might call a 'thin' conception of republicanism to a thicker green conception. Republicans argue that freedom entails participation, but green republicans morph this so that freedom entails agreement. Participation is reduced to a particular style of participation and the citizen is expected to be a particular type of citizen. This becomes clearer if we consider the virtues that the green republican citizen is expected to display.

Eco-Virtues

Green republican citizens are outward-looking, virtuous members of the community who actively participate in the public realm and are sensitive to the long-term well-being of the environment. We need further clarification here about what constitutes the duties and virtues of green republican citizenship. Barry, following various republican arguments, seems to demand of green republican citizens only the duty to participate actively in community life. It is this participation that ensures individual freedom. However, he also notes that responsible citizenship cannot be reduced to participation: 'simply increasing the participation of citizens in democratic decision making is no guarantee that they will act responsibly'.[40] As Kymlicka and Norman put it: 'emphasising participation does not yet explain how to ensure that citizens participate responsibly – that is, in a public-spirited, rather than self-interested or prejudiced way'.[41] Green republicans rely upon the notion of civic virtue to ensure that citizens participate responsibly, but the confusing problem with this conception of civic virtue is that, as well as being the condition of active citizenship, it also seems only to arise through that active citizenship. Curry states exactly this: he defines ecological virtue as 'both the practices that encourage the qualities of citizens, and the qualities that enable those practices, without which the common good, either as reality or ideal, cannot flourish or indeed long survive'.[42]

Curry leaves it open as to how such virtue is cultivated, which leaves us with little inkling as to how his green republicanism might be implemented. But Barry is aware that it is all too easy for the obligations of citizenship to be forgotten or ignored: 'The republican conception of politics and citizenship is explicit in recognising the ever-present temptation for citizens to forget their duty or lapse into self-regarding interests and pursuits at the expense of fulfilling their individual contribution to collective action and vigilance for the common good.'[43] Barry argues, therefore, that it is the role of the state to promote the conditions for green republicanism. He advocates a 'compulsory sustainability service', a state scheme in which all citizens spend a certain amount of time cleaning up beaches or working on community farms.[44] In other words, citizens should be forced to participate in the social and environmental common good. Barry argues that this sort of scheme should not be seen as an authoritarian demand for compliance, but rather as part of the struggle against the underlying causes of environmental harm. Here he states: 'if one accepts the argument for sustainability, one does not just have the *right* to demand changes to create a more sustainable society but one also has the *obligation* to do so'.[45] But this seems to be a perfect example of a hegemonic substantive conception of a common good overriding alternative conceptions of the good life.

James Connelly takes the cultivation of eco-virtues in the virtuous citizen further. He explains that a virtue is a disposition to act in a certain way, and that these dispositions are key in instituting sustainability; externally motivated environmentally friendly actions are necessary but are not sufficient; virtuous citizens internalise the aim and value of good environmental actions: 'Standard policy instruments and economic incentives are not in themselves sufficient to achieve sustainability. We also need to change environmental dispositions and habits.'[46] Connelly agrees that eco-virtues must be nurtured by the state: 'Being virtuous precedes virtuous being.'[47] He explains that 'because responses to the environment have to go all the way down, environmentally sensitive dispositions need to be developed and encouraged'.[48]

Attentiveness to eco-virtues is certainly a useful approach for greens because (as we have seen in previous chapters) it is hard to assign blame for an environmental issue such as climate change, where the causes are spread out in both time and space. Advocates of eco-virtues suggest that rather than assigning blame for actions, we should think about the attitudes we would like to encourage in people as the yeast in fermenting a more responsible – and sustainable – world. As Thomas E. Hill states, it is hard to find convincing reasons why environmentally destructive acts are wrong.[49] Appealing to the rights or interests of non-human nature, for example, is notoriously difficult. But by attending to virtues, rather than actions, we can offer a convincing and positive account of the traits necessary for an environmentally friendly society. One of the virtues named by Hill is humility, which encourages individuals to value things other than themselves and their immediate circle of friends and family. The virtues mentioned by other theorists include courage, frugality, prudence, patience, compassion, charity and asceticism.[50]

Eco-virtues are, for Connelly, not directed towards the Aristotelian *eudaemonia*. They are not about achieving good for the individual, but are directed outwards towards achieving the environmental good – Connelly is interested in what he terms 'dirty virtues', and if well-being and happiness are also achieved then this, he says, is simply a bonus: 'Virtues go beyond their bearers; they are not private but social, and their exercise therefore requires a conception of the common good.'[51] One of the key eco-virtues, he agrees, is exercised in participating in the discussion about the common good.[52] He states that the common good is not pre-given and that citizens need to reflect upon and debate their own dispositions and duties: 'Ecological citizenship comprises the ecological duties together with the virtues appropriate to their fulfilment. This includes the duty of deliberating on duties: we have a duty to ask what our duty is.'[53] To come to a decision on what our duties are, it is important to note, we need the virtue of rational deliberation: 'An eco-virtue is an internally motivated ecological thoughtfulness leading to action. The virtue of rational deliberation … is essential to the proper formulation and understanding of our eco-duties.'[54] Conceptions of eco-virtues again rest upon the notion of the 'rational individual' who is able to come to agreement. Connelly states that just because we don't completely agree doesn't mean we completely disagree. But is 'not completely disagreeing' enough to produce a similar account of the responsibilities expected of the citizen and the role of the state in guaranteeing the fulfilling of the responsibilities and encouraging the development of eco-virtues? I contend that this depiction – of the rational, virtuous, citizen who freely participates in a discussion and arrives at an agreement about what the common good means

– leaves out something important. It forgets that there are many different ways to be and to think. It is difficult to reconcile the assertion of the notion of eco-virtuous citizens with an appreciation of the diversity of human societies. Societies contain a myriad of ethical values and duties, and a multiplicity of ways of perceiving 'nature'.

For green republicanism, one of the key virtues of the citizen is rational thinking, and one of the key steps in responding to the issues of climate change is rational agreement upon the common good. We have arrived back at the rational individual whom we met in the previous chapters. Addressing climate change, again, appears to require the human being to think and act in the correct way. What about those who disagree, or who are already excluded from participation because they do not exhibit the virtues of green republicanism and fail to fulfil its responsibilities?

Gender and Responsibility

The 'turn to citizenship' in green thinking demands an investigation from a feminist perspective. In Aristotle's city-state, citizenship was limited to 'men of leisure'. As one commentator writes: 'His definition of citizen includes only a small part of the population of any Greek city. He is forced to admit that the state is not possible without the cooperation of men whom he will not admit to membership in it.'[55] Doubly excised – marginalised by Aristotle but forgotten in this quote too – are the women, who are excluded from the public sphere and confined in the private. The question posed by contemporary feminism concerns how much has changed? The very concept of citizenship is accused by some feminists of continuing to manifest a masculinist bias. Citizenship is one of the concepts Anne Phillips is referring to when she states that 'feminists have queried most of the basic concepts of political thinking, arguing that theorists have always built on assumptions about women and men, though they have not always admitted (even to themselves) what these were'.[56] Despite claims that citizenship is now a universal category, giving every member of a community an equal status regardless of race, gender, class and so on, this assumption of a generic model masks the inevitable exclusions of a category arising from such a particular conception. The tendency is, as Phillips puts it 'to smuggle real live men into the seemingly abstract and innocent universals that nourish political thought'.[57]

While citizenship might give a formal, equal legal status, there are various political, economic, cultural and social barriers preventing certain sections of the population from participating equally as citizens in the public realm. Thus 'one disadvantaged group after another has fought lengthy battles to be included on the list of people entitled to citizenship, only to find that social justice and equality still eluded them'.[58] Although rights may be universally extended to all at least in theory, with the shift to responsibility demanded by green republicans, the problem deepens. Is everyone equally able and equally equipped to be an active and virtuous citizen and to fulfil their citizenly duties? And how equal are these duties? Are some people burdened more greatly by them?

Sherilyn MacGregor notes that much green citizenship literature ignores these issues. She points out that republican conceptions demand that individuals give up their free time to participate in the public sphere. And yet not everyone has free time to give up: 'the image of environmental citizens as simultaneously self-reliant and politically active is highly problematic from a feminist perspective'.[59] The very existence of free time depends upon the supportive work provided, usually, by women. As Ruth Lister writes: 'The question of time is a crucial one. Implicit in the notion of active citizenship is an assumption about time: that people have time to be active citizens, be it as good neighbours or volunteers or as active participants in the political life of the community. This assumption is rarely made explicit. Once it is, it raises important questions about the distribution of time between men and women.'[60] MacGregor queries whether theories of green citizenship even notice this issue. 'Green political theorists have recognised human dependency on natural processes.... Most, however, have failed to acknowledge human dependency on the caring services performed by human beings, and thus have failed to value the role played by domestic life in the search for ecological sustainability.'[61] And as she goes on to point out, those that fail to fulfil the responsibilities of citizenship are ultimately dismissed as 'irresponsible'.

It must be questioned not just whether everyone will be equally able to discharge the responsibilities of green republicanism, but whether those responsibilities are distributed equally. While on the one hand the primary republican responsibility is to participate in politics, there are also duties arising from the need to live sustainably and to mitigate and adapt to climate change. But the content of these latter duties are decided of course by those 'men of leisure' conversing in the public sphere. The move to a more sustainable society demands an intensification of labour, and that this labour is divided unequally. While the green republican citizen participates in debates about the duties of citizenship, someone is minding the children, taking out the recycling and washing the reusable nappies. We cannot make the gendered assumption that this someone is necessarily a woman, but the point remains that green republicanism must be careful not to gloss over the existence and vital role of the domestic sphere: 'Communities, families, and the women who care for them have little choice but to shoulder the ever-greater burdens of responsibilities being created.'[62]

Feminist work draws attention to the gender divisions of labour in coping with climate change and how the predicted increase in serious health issues is likely to increase the burden upon women, since it is they who are likely to be the carers for the sick and elderly.[63] We should be careful not to generalise and essentialise women's roles, but, as MacGregor points out, conceptions of green citizenship often produce and reproduce gender stereotypes and constructions of hegemonic femininity and masculinity. She analyses the existence of the 'EcoMom Alliance' and the emergence of what she terms 'eco-maternalism': 'women's maternal role is often used as a justification for their involvement in environmentalism'.[64]

The gender-differentiated impact of climate change is now increasingly well documented. As Geraldine Terry notes, it is particularly the vulnerability of women living in rural areas of the developing world that is acknowledged.[65] This, however, does not give us the whole picture. It is not just the impact of climate change upon individuals as victims, but also the changing conception of obligations and citizenship that it feeds into, that is gendered. In his description of green republicanism, Barry states that 'republicanism emphasises the importance of active citizens'.[66] But who, exactly, are these active citizens? Who is expected and able to enact his compulsory sustainability service? Who is excluded and overlooked in republican citizenship? Although I have focused upon women in this chapter, other groups within a population are also sidelined. Bronwyn Hayward points out how children, too, are omitted from discussions about citizenship and the environment.[67] Feminist analysis of citizenship and climate change importantly highlights the various presuppositions and exclusions built into the green republican perspective. Such analysis reaffirms my argument that green republicanism relies upon a particular depiction of the individual and brushes over differences and disagreements in today's plural societies.

Good and Green

Many, if not most, of those concerned by climate change despair at the politicians' feeble responses to global warming. As one newspaper columnist laments, 'if we want the planet to be saved ... people have to be persuaded to make sacrifices for the common good' but the problem is 'the common good has become, to Western politicians of both left and right, a completely alien concept. They have no language in which to convey to their electorates the importance and urgency of what needs to be done.'[68] Green republicanism attempts to provide the words in which to reclaim the notion of the common good, rewriting them in bold green marker pen. In doing so it offers an enriched and positive vision of community in which concerns about environmental issues such as climate change are incorporated within a wider awareness of the common good. It reverses the individualised approach of previous chapters to consider the role of the community. As part of a collective the individual is able to participate in a coordinated effort in tackling climate change.

Ultimately, however, the green republican approach ends up enforcing one green vision over others. It asserts the possibility of consensus on the issues of climate change and asserts a particular model of the citizen. It is vital to acknowledge that not everyone in society fits this model. The rights and responsibilities of citizenship are not equally and uniformly distributed. Feminist analysis reveals the differentiated burdening of citizenship.[69] Some have greater access to and manoeuvrability within the discussion that determines the content of the common good, while others are expected to enact this content to which they have had little chance to dissent. Is there a way to act as a collective but respect difference? These questions of democratic participation, inclusion and exclusion bring us to the deliberative democratic approach to climate change, the topic for the next chapter.

4 Beyond Conflict?
The Green Deliberative Democratic Approach

Until philosophers are kings, or the kings and princes of this world have the spirit and power of philosophy, and political greatness and wisdom meet in one, and those commoner natures who pursue either to the exclusion of the other are compelled to stand aside, cities will never have rest from their evils – no, nor the human race, as I believe – and then only will this our State have a possibility of life and behold the light of day. – Plato, *The Republic.*[1]

Who should rule, Plato invites us to consider: the unruly multitude who bay for power or the wise and benign expert upon whom power is involuntarily thrust? In the place where democrats argue no one should sit as the sole and permanent wielder of sovereignty, Plato enthrones the philosopher. For Plato, ruling is an art in which only a few have the necessary wisdom, and rule by the demos would therefore descend inexorably into chaos. He is surely right that we would prefer to entrust the sailing of a ship to the hands of a learned captain rather than be navigated by its motley crew, so why doesn't the same hold for the planet earth and its climate? Why, today, should the term democracy – for a long time a disparaged notion – be held in any high regard, not least by the environmentalists whose theories we examine here?

The previous chapter challenged the green republican assertion of one commonly accepted notion of 'the good' consisting of a particular vision of a sustainable, climate-friendly way of life. The problem with this solitary green vision is that it enforces a particular set of responsibilities and virtues upon its citizens, blotting out the diversity of human society and forgetting many who are excluded from political discussion. Yet the question that has not yet been addressed is *why this should matter*. Environmentalists might argue that, while regrettable, a loss of democratic differences should not deter us from pursuing the environmental good and implementing solutions to tackle climate change. Why should democracy be valued above saving the lives of millions, particularly of those more at risk? Isn't democracy a luxury in such situations of crisis?

This chapter begins therefore by asking why, in addition to the intrinsic value it holds, democracy is actually important for adequate responses to climate change and why eco-authoritarian approaches, despite superficial appearances to the contrary, hinder such responses. It then considers in more depth the green deliberative democratic approach which, importantly, connects democracy to the resolution of climate change. However, I argue that, while breaking new ground, the green deliberative democratic approach ultimately fails because it assumes that any response to climate change rests upon reaching agreement. Deliberative democracy doesn't rely upon any substantive notion of 'the good life', it rather celebrates the myriad of differences in society. Nevertheless, by attempting to bring these differences into alignment, through the exercise of rational communication, it cannot be as inclusive as it pretends. I suggest that the notion of 'communicative rationality' excludes alternative ways of thinking and that the goal of consensus marginalises difference.

Although ostensibly very opposed, both eco-authoritarian and deliberative democratic approaches presuppose that action against climate change in a democracy hinges upon agreement; they just differ on whether this agreement is a feasible goal. As I will go on to show in Chapter 5, the assumption that action against climate change rests upon agreement can be challenged. I will argue that it is, in fact, from political disagreement that collective decisions emerge. For now, however, I will focus on the reasons why climate change demands a democratic response.

Problems of Democracy

I have argued throughout the previous chapters that a *collective* response is necessary in addressing the issue of climate change. This raises the question of where the collective support for climate change policies can actually be located.[2] Many, such as Polly Toynbee, are sceptical that there is any public will to implement climate change policies: 'Fixing the climate is not a purely practical conundrum, it is a purely political problem. We could build the windmills, the solar, the nuclear and whatever it takes to be self-sustaining with clean energy for ever if we wanted to. But enough people have to want to change how they live and spend to make it happen. So far they don't, not by a long chalk.'[3] Climate change is not among the voting priorities for most of us; recent research suggests that only 17 per cent of UK voters in the general election in 2010 listed climate change as one of their top three or four issues; climate change appears to have minimal 'shoving power'.[4]

In a Policy Network paper, the well-known political theorist Anthony Giddens acknowledges the seemingly inevitable difficulty in addressing long-term environmental issues within democratic political systems: 'By their very nature ... democratic countries tend to be driven by the immediate concerns of voters at any one time.' And he asks, 'How do we think long-term in societies that tend to be dominated by short-term issues?'[5] Giddens's answer is to institute committees and agreements that transcend political divisions: 'Forging and sustaining a cross-party consensus on climate change policy would help a great deal with being able to take a long-term view in policy issues.'[6] Giddens seems to give up on democracy, and is resigned to banishing the possibilities of political disagreement. He suggests that the clashes between political adversaries result in a deadlock from which no climate change policy can emerge: 'An adversarial political system is difficult to reconcile with long-term thinking, since where needed climate change policies are unpopular, a party might simply surrender to populism in search of political advantage.'[7] Hence Giddens prefers the creation of an 'ensuring state' as opposed to a weaker 'enabling state'. An ensuring state is one that 'is expected or obligated to make sure such processes achieve certain defined outcomes'.[8] In his recent book on climate change he extends these ideas, stating that 'there should be an agreement among competing political parties that climate change and energy policy will be sustained in spite of other differences and conflicts that exist'.[9] Following such a plan, however, takes us away from democratic politics into the dubious territory of the deliberate misleading or marginalisation of the public.

Indeed, another paper from the Policy Network states quite explicitly that governments need to employ strategies that actually bypass the public and that hit hardest those sections of the population who have least political voice and clout. It counsels 'policies that target losses on small sections of society, particularly groups that are least able to inflict political damage via the ballot box or to exercise threats to withdraw investment from the country'.[10] This statement blatantly advocates an undemocratic and socially unjust tactic. Is this the only option in tackling climate change? Dougald Hine acknowledges the necessity for greater austerity, but rejects the idea that it must be imposed by an authoritarian state. He writes that 'mechanisms for restricting our personal behaviour will be required, but we should demand involvement in the process rather than petitioning the state to relieve us of our freedom'.[11]

The assumption of many theorists and policy makers seems to be that entrusting climate change to democratic decision making is dangerous, because it is likely to produce the 'wrong' decision. Democratic politics is regarded as an impediment to ensuring collective decisive action to tackle climate change. Many see the only option as an 'eco-authoritarianism' that coerces a sulky dissenting public for its own good. Eco-authoritarianism was a popular strategy in the 1970s, when theorists such as Garrett Hardin posited the necessity for coercion to curb exponential growth of the human population.[12] This sort of strategy has had a revival recently in response to climate change.[13] I will examine, next, arguments of eco-authoritarianism and argue that far from being the sole political route to solving climate change, such anti-democratic measures are likely to impede movements towards helping the environment.

Eco-Authoritarianism

Theorists and policy makers concerned about climate change often despair of democratic politics. James Lovelock, for example, doubts that modern democracies will ever be able to respond adequately to the urgent problem of 'global heating': 'We need a more authoritarian world. We've become a sort of cheeky, egalitarian world where everyone can have their say ... it may be necessary to put democracy on hold for a while.'[14] He proposes an eco-authoritarianism in which a 'few people' are accountable, but are able to enforce the measures necessary. 'If our leaders were all great and powerful they could ban the keeping of pets and livestock, make a vegetarian diet compulsory, and fund a huge programme of food synthesis', but 'orderly survival requires an unusual degree of human understanding and leadership and may require, as in war, the suspension of democratic government'.[15]

David Shearman and Joseph Wayne Smith offer a detailed description of eco-authoritarianism. They argue that climate change will never be solved democratically: 'Each year the certainty of the science has increased, yet we have failed to act appropriately to the threat',[16] they point out, explaining that individuals are never going to vote for a reduced standard of living, leading them to the conclusion that 'democracy itself is a big problem'.[17] They advocate a Platonic form of authoritarianism in which the rule of experts replaces the perils of democracy.

In a paper entitled 'The coming of environmental authoritarianism', Mark Beeson argues that authoritarianism is likely to be consolidated in regions of the world that are particularly vulnerable to climate change and which already have a propensity towards that form of regime. He insists that democracy is antithetical to environmental sustainability, and ponders whether authoritarianism will reach parts of the world with longstanding traditions of democracy: 'forms of "good" authoritarianism, in which environmentally unsustainable forms of behaviour are simply forbidden, may become not only justifiable, but essential for the survival of humanity in anything approaching a civilised form'.[18]

Others, however, have used empirical research to suggest that democracy is correlated, loosely, with a fall in environmental degradation.[19] There are various reasons cited for this: by allowing scientists to research, travel and exchange their ideas freely, democracies encourage the advance of science.[20] Democracies ensure greater transparency and media freedom, which make it more likely that environmental problems are reported and people are more likely to be informed about climate change. Climate change is also more likely to appear on the political agenda in democracies, since there are more channels through which individuals and groups can voice their concerns. Whereas in a totalitarian system power is closed off from new influences and interests, in a democratic system, green parties are able to rise to power.[21] Environmental activists are less likely to be harassed or imprisoned. Some argue that human rights are better respected in a democracy, and therefore a democratic government is more likely to respond to environmental problems that are life-threatening to its population. Finally, there is the claim that there is a correlation between wars and environmental damage, and that democratic states are less likely to engage in war.[22] Summarising many of these points, Rodger Payne writes: 'Thanks to free speech, free press and other individual liberties, it is possible for the combined forces of the mass media, various environmentalist movements and relevant scientific communities to monitor the activities of the most prominent sources of environmental degradation – often corporations and governments – and publicise their findings, however critical.'[23]

Yet, despite all these claims, there is no undisputed clear link between democracy and environment; the benefit of a democratic system in tackling an issue such as climate change is difficult to establish empirically. As Eric Neumayer writes rather unhelpfully, 'while a good theoretical case can be made for a positive link between democracy and environment, there are a number of considerations pointing in the opposite direction'.[24] The empirical studies assessing the impact of democratic systems on the environment rest upon what 'democracy' is understood to mean; they therefore are already substantively underpinned by a theoretical account of democracy. But 'democracy' is very difficult if not actually impossible to define.[25] Many of the arguments above presuppose that democracy is a particular, liberal and representative version of democracy in which the liberty of the individual and of the market is guaranteed, and in which individuals are only minimally involved in politics.

By reconsidering the meaning of democracy it is possible to see the apparently inherent problems of democracy for the environment in a different light. For Val Plumwood, the failure of modern democracies to address environmental problems, despite a general awareness of them, is actually a failure not of democracy but of the current political system – that is actually not very democratic at all: 'the failure of ecological responsiveness where there is widespread citizen concern and support for change, can be taken as an indication that something is rotten, that communication and participation are somehow blocked or skewed'.[26] It is a mistake, she points out, to assume that the fault lies with a population of self-interested consumers who are too preoccupied with personal goals. She explains that this analysis is convenient because it absolves liberal democracy from responsibility. But it can only do so if it is presupposed that humans are essentially self-maximising, ecologically harmful consumers. Plumwood sees the alternative of democracy or the environment as a false choice: 'it is not democracy that has failed ecology, but rather liberal democracy that has failed both democracy and ecology'.[27] It is actually unsurprising that environmental concerns are excluded from a system in which a privileged elite are able to insulate themselves from the impact of environmental destruction, have the greatest interest in the continuation of the status quo, and are the very people in control of decision making.

Following her work, we see that it might not be democracy, *per se*, that is a problem for tackling environmental problems, but rather a particular undemocratic version of it that shores up the power, conflict and exclusion in the political realm, and hinders the real political interaction of the demos. Relying on elites and experts to solve the problem will not work because they, by definition, have less to gain by doing so. As Bob Pepperman Taylor argues, 'if policy is defined and controlled solely by experts, elites, ideological minorities or philosopher kings it necessarily represents the interests, concerns and values of only a fraction of the community'.[28] Rather than a rejection of democracy, what is needed is *more* democracy. By opening up the discussion to new voices and more democratic difference, the lethargy that is a problem in tackling climate change can itself be tackled. This leads us to examine, next, recent re-visions of democracy that aim at including everyone within the discussion, and see this rejuvenated democracy as a potent new way of approaching climate change.

These visions of democracy, however, have problems of their own. As Plumwood notes, simply widening the scope of participation is not enough. Using 'critical ecological feminist' analysis that problematises the division between a public sphere of reason and a private sphere of nature, Plumwood explains that what is needed is a recognition of the relations of power that exist across the various social arenas and result in interlinked oppressions.[29] Do the new visions of democracy do enough to highlight the existing exclusions and entrenched power that prevent a multivocal response to environmental problems?

The Deliberative Turn

David Held and Angus Fane Hervey consider the problem for governments in acting on climate change. As they put it, 'it is extremely difficult for governments to impose large-scale changes on an electorate whose votes they depend upon'.[30] Yet they nevertheless reject an 'eco-authoritarian' solution and argue that authoritarian governments have less incentive to adopt or stick to environmentally sustainable policies.[31] This does not mean, they are careful to warn, that our current democratic systems are able to handle the problem of climate change. Held and Hervey advocate the evolution of democracy towards a new deliberative form. The model of 'deliberative democracy' is often described in comparison to conventional 'aggregative democracy'. Aggregative democracy acknowledges the existence of different preferences in a society and suggests that these can be handled in a democratic system by voting, or by arriving at a compromise, and thus 'aggregating them'. Individuals are seen as possessors of interests that are fixed prior to any political discussion, and this political discussion is seen as consisting of self-interested and instrumental bargaining. But this sort of democracy doesn't seem to encourage the interaction and inclusion of marginal and minority voices.

Deliberative democracy, however, is resolute in its aim to welcome difference. By bringing the different perspectives in a society into deliberation together, this model of democracy seems to be much more inclusive and legitimate than other models. For Taylor, 'we cannot begin to understand who we are as a people, we cannot have a principled public life on environmental or any other issues, unless public policy grows directly out of democratic deliberation'.[32] The 'deliberative turn' has altered the very source from which democracy finds its validity: 'The essence of democracy itself is now widely taken to be deliberation, as opposed to voting, interest aggregation, constitutional rights, or even self-government'.[33] Deliberative democracy is now the new orthodoxy.[34]

One of the most influential theorists of deliberative democracy is Jurgen Habermas, whose work seems to hold the recipe for combining respect for difference with inclusive discussion. He acknowledges that in modern plural societies different cultural forms of life exist with different substantive ethical values. Yet everyone, he believes, regardless of their different ethical perspectives, can be integrated into an inclusive political culture that rests upon shared moral and constitutional principles: 'in complex societies the citizenry as a whole can no longer be held together by a substantive consensus on values but only by a consensus on the procedures for the legitimate enactment of laws and the legitimate exercise of power'.[35] Unlike a republican model of democracy, there is no common set of 'thick' ethical principles held by the entire community. But, unlike a liberal model, democratic discussion is not just a matter of finding a compromise between competing interests.[36] Habermas crucially emphasises the role of the democratic procedure itself in creating an overarching unity and set of shared moral values. He offers a vision of democracy in which different and competing perspectives become part of the discussion. It is not through any prior cultural, national or ethnic identity, but through the debate itself, that people are brought together: 'deliberative democracy … no longer hinges on the assumption of macro-subjects like "the people" of "the" community'.[37] The discussion does not exclude anyone who has a relevant contribution to make but is fully inclusive: 'new forms of life are imported which expand and multiply the perspective of all'.[38]

Habermas, however, relies on a particular vision of democratic discussion. The reason it can incorporate all differences is that it is guided by 'communicative rationality'. As Seyla Benhabib explains, deliberation involves the articulation of ideas in public, a procedure that must be supported by *good reasons*. A 'reflexivity' is introduced into the discussion which compels the deliberators to think from a less partial perspective: 'This process of articulating good reasons in public forces the individual to think of what would count as a good reason for all others involved.'[39] Whereas the aggregative model of democracy features an instrumental form of rationality, the deliberative model hinges upon a communicative rationality through which arguments will rise above the interlocutor's particular perspective: 'argumentation of its very nature points beyond all particular forms of life'.[40]

Through communicative rationality, interlocutors can consider each other's positions and perspectives. While deliberative democrats acknowledge the existence of differences in preferences they do not, like aggregative democrats, regard them as irrevocably irreconcilable. Benhabib explains that individuals should not be presumed to have already formulated and fixed ideas, but rather to have mutable preferences that, by being exposed to one another, can be transformed: 'it is incoherent to assume that individuals can start a process of public deliberation with a level of conceptual clarity about their choices and preferences that can actually result only from a successful process of deliberation'.[41] In a deliberative democracy, no particular perspective can claim to be the 'right' one from the outset, but rather must justify itself through public debate: 'deliberative democrats advocate that democracy moves away from any notion of fixed and given preferences, to be replaced with a view that democracy should become a learning process in and through which people come to terms with the range of issues they need to understand in order to hold defensible positions'.[42] Deliberative democracy suggests that, by being brought into contact with each other in a rational discussion, apparently conflictual perspectives can be aligned and brought to an enlightened choice: 'the key objective is the transformation of private preferences via a process of deliberation into positions that can withstand public scrutiny and test'.[43]

Inherent in the deliberative process, it is important to note, is the goal of reaching agreement and it is this *telos* of agreement that is expected to ensure that the discussion is rational and impartial: 'the practice of argumentation sets in motion a cooperative competition for the better argument, where the orientation to the goal of a communicatively reached agreement unites the participants from the outset'.[44] Some deliberative democrats query whether deliberation is necessarily set upon reaching full consensus. For example, Graham Smith suggests that 'rather than consensus, democratic deliberation is best understood as being orientated towards *mutual understanding*, which does not mean that people will always agree'.[45] Yet Smith still argues that the aim of deliberative democracy is to resolve conflict through inclusive dialogue that is aimed at reaching shared ideas: 'democratic deliberation encourages mutual recognition and respect and is orientated toward shared understanding and the public recognition of the common good'[46] and this 'common good' presumably consists of some sort of shared substantive vision.

Through democratic deliberation, underpinned by communicative rationality and guided by the telos of reaching agreement, individuals with different cultures and conceptions of the good life are included in the discussion and aim at reaching an impartial and rational agreement that is validated by its deliberation and is acceptable to and accepted by all. Although any agreement is always open to challenge, it is expected to hold and wield legitimacy at least temporarily.

Green Deliberative Democracy

For greens, the notion of a communication that includes differences but potentially results in a social solidarity and agreement on environmental policy is clearly of great value. Deliberative democracy seems to hold the answer to overcoming the reluctance of a population to tackle climate change. Advocates of this approach suggest that, through debate in the political realm, those with very different perspectives about the problem of climate change, through a mutual adjustment of ideas, can be brought to a greener and more rational conclusion acceptable to all them.

Note that deliberative democrats do not posit the existence of one singular debate occurring within an open public arena but rather a 'plurality of modes of association' which include political parties, social movements, citizen initiatives, voluntary associations and consciousness-raising groups.[47] This description of plurality neatly fits the myriad of different identities and associations centred around the term climate change.

John Barry recommends a deliberative democratic element as part of his vision of green republicanism, encountered in the last chapter. In order to reach agreement upon the common good, Barry recommends deliberative democracy in which citizens educate each other: 'the pedagogic nature of deliberative democracy is not about the internalisation by the populace of some given truth as determined by experts. Rather, the pedagogic effects of deliberative democracy is a process of mutual learning.'[48] Although there is no predetermined common good, the result of deliberative democracy is a move to an agreement on its meaning, an agreement that due to public reasoning is necessarily one that is understood by Barry to be rational: 'The introduction of communicative rationality to the coordination of individual action makes it less likely that the collective result will be ecologically irrational.'[49]

Smith, too, promotes deliberative democracy for the opportunities it provides for green perspectives. 'For Greens,' he writes, 'the moralising effect of deliberation offers the opportunity to emphasise the ethical nature of human–non-human relations and the public good character of many environmental problems and to expose and challenge the narrowly self-interested grounds of many environmentally degrading and unsustainable practices.'[50] He warns, however, that there is no guarantee of an environmentally friendly outcome – but nevertheless holds that deliberative democracy creates a forum in which greens can at least present their concerns. For Smith, deliberative democracy enhances reflection on the myriad of different environmental values, improves participants' perspectives, and thus legitimises environmental policy decisions.

John Dryzek presents a stronger case for deliberative democracy producing a green result. Dryzek is perhaps the best-known advocate of deliberative democracy (or his particular interpretation of it, which he calls 'discursive democracy') in addressing environmental issues. Whereas today's liberal aggregative democracy is focused upon particular economic interests, the deliberative version is better, he explains, at incorporating the general ecological interest. He therefore extends deliberative democracy in a green direction by including the communications not just of diverse human beings but of the natural world, too. He thus suggests that we 'look for the essence of democracy not in the aggregation of interests or preferences of a well-defined and well-bounded group of people ... but rather in the content and style of interactions'.[51] Whereas non-human entities do not have preferences exactly, they do communicate in various ways, through ecological processes such as cycles, the creation of niches and through feedback signals. Dryzek urges us to battle our human autism and pay attention to the communication emerging from nature: 'authentic deliberation involves reflection upon preferences induced by communication in non-coercive fashion. There is no reason why this communication has to have a human source.'[52] His 'ecological democratisation' suggests that listening – rather than speaking – is the key political practice and that the boundary between the human and non-human should be transcended so that nature is heard: 'recognition of agency in nature therefore means that we should listen to signals emanating from the natural world with the same sort of respect we accord communication emanating from human subjects, and as requiring equally careful communication ... communicative interaction with the natural world can and should be an eminently rational affair'.[53]

These theorists all believe, then, that deliberative democracy is an important tool for tackling environmental problems. For Barry, participating in the discussion will educate citizens and transform them into ecological stewards. For Smith, deliberative democracy offers a space in which environmental values can hope to transform conceptions of the common good. For Dryzek, the deliberative version of democracy can incorporate nature in a way that other versions of democracy cannot, and therefore is more likely to produce a green outcome. However, there are weaknesses in this strategy that the rest of this chapter is devoted to exposing. The participants within a deliberative democracy are regarded as free and equal, and capable of presenting their ideas in a 'rational' way. To quote again from Benhabib, 'according to the deliberative model of democracy, it is a necessary condition for attaining legitimacy and rationality with regard to collective decision-making processes in a polity, that the institutions of this polity are so arranged that what is considered in the common interests of all results from processes of collective deliberation conducted rationally and fairly among free and equal individuals'.[54] But these claims of deliberative democracy – to be able to move through an inclusive debate guided by 'communicative rationality' to a green outcome on an issue such as climate change – are highly problematic. After closer examination of 'communicative rationality' I argue that the deliberative democratic depiction of the discussion as free, equal and rational is misguided. I claim that although the discussion may well be opened up to others previously excluded, there are still inevitable exclusions right from the outset. Secondly, I argue that the ultimate aim of green deliberative democracy to arrive at an inclusive agreement on climate change is an *impossibility*, and that this ignores the inevitability of power relations in political discussion. It therefore undermines its own claims for the inclusion of differences.

Rationality and Otherness

Iris Young provides an important objection to deliberative democracy. She argues that many deliberative democrats assume that by removing economic and political power from a deliberative forum, the speakers there will become free and equal, but they forget the social power that exists. Various cultural and social differences mean that speakers have very different ways of speaking and different senses of entitlement: 'even communication situations that bracket the direct influence of economic or political inequality nevertheless can privilege certain cultural styles and values'.[55] The style of deliberation that is taken to be neutral is actually culturally specific and marginalises other ways of communicating: 'Despite the claim of deliberative forms of orderly meetings to express pure universal reason, the norms of deliberation are culturally specific and often operate as forms of power that silence or devalue the speech of some people.'[56] Young suggests, therefore, that other forms of communicating should be included in the deliberative forum and she suggests storytelling, greeting and rhetoric. These different ways of speaking will open up the debate to those currently excluded.

Dryzek takes on board Young's complaints. 'Deliberative democracy', he claims, 'need not fear difference.'[57] He claims that deliberative democracy can overcome the challenge that it relies upon a particular type of communication that privileges a certain kind of person by admitting, conditionally, other forms of communication. Feedback signals from nature, he suggests, are one such alternative form of communication that can help deliberative democracy be attuned to plurality. Yet, as he points out, there is no guarantee that these other forms of communication won't also be coercive and excluding. His solution is to suggest, therefore, that all forms of communication used in deliberation are subject to 'standards of communicative rationality'.[58] But this just reproduces the problem; he attempts to shoehorn these different ways of using language into one supposedly 'rational' and neutral discussion. The communications from nature must be interpreted and somehow brought into reasoned agreement with all other perspectives. Dryzek (and Young too) assume that exclusion can be excluded from the discussion.

In her critique of Habermas, Diana Coole questions the idea that individuals are fully self-aware and rational and that they are capable of communicating directly and clearly. Such an assumption ignores how the human individual is subject to unconscious motivations and strangeness to themselves: 'Habermas gives subjects privileged communicative access to their inner selves such that, despite communicative blockages, there is in principle nothing immune to rational self-communication.'[59] He assumes that our use of language can allow entirely transparent communication. Habermas therefore dismisses otherness (or what Coole calls 'alterity') as the residue of some traditional primitive culture; something 'historically obsolete, mystical and unverifiable'; something *external* to language and reason, which can be overcome through rationalisation. 'For Habermas, failings in communication can only signify a distortion to be overcome, never a salutary challenge to the hubris of reason.'[60]

But, for Coole, otherness cannot simply be transcended through communication, for otherness is actually *internal* to language and culture. There is a limit to the ability of language to render the world transparent. Beneath ostensible rational discussion operates the non-rational and the pre-discursive. It is this irreducible alterity in language itself that Habermas tries to suppress. But this suppression is an act of violence, and to deny that it occurs is to be blind to the power relations that operate at the pre-discursive level and affect our abilities to participate freely, equally and rationally: 'Habermas largely ignores the non-systemic, nondiscursive processes of power and politics that continue to circulate within the lifeworld even in modernity.'[61]

Coole goes on to assert the importance of a 'politics of alterity', which is alert to the exclusions and silencing that goes on through supposedly free and equal negotiations of democratic discussion. Her work points to the difficulties for a deliberative democracy that is supposed to arrive at an inclusive agreement on a policy about climate change. We are never fully aware of our own unconscious motivations and wishes and this must impede our ability to communicate rationally: 'ordinary language is intrinsically riven by slippages, metaphors, absences, deferrals, desire and change associations which mean that it always communicates both more and less than conversationalists intend'.[62] Aren't distortions of mutual understanding impossible to expunge from communication? Don't interlocutors stumble over slips of the tongue? Don't we say too much without meaning to? Don't we often fail to express what we mean, and don't our fellow discussants hear something else entirely? Coole demands that we consider how our choices and ideas are conditioned prior to political discussion and are reproduced within that discussion. Body language, for example, can dismiss some, and empower others. Can deliberative democracy ever eradicate such exclusions? The climate talks in Bangkok betray such confusing workings of language: 'Even seasoned diplomats find the talks are surreal, with an arcane language, logic and a pace of their own', writes John Vidal, while 'negotiators admit to becoming lost in their own verbiage … the talks have invented their own language. There are Bingos … who discuss Mrvs … Namas ….'[63]

This critique raises particular questions for Dryzek who, as we saw above, wants to include the communications of nature within his discursive democracy. He shifts emphasis away from speaking and towards hearing; as participants in a discursive democracy we are supposed to *listen;* to open our ears more than our mouths. Nature is construed as humans' 'other', and its otherness is something to which Dryzek urges us to pay attention. This strategy may appear to chime with Coole's project, but while Dryzek attempts to incorporate the otherness of nature into the debate, he also aims at rendering it 'rational'. His aim is to render this otherness transparent within communicative rationality; he seeks to rescue communicative rationality from Habermas and to create a 'more egalitarian interchange at the human/nature boundary'.[64] But who is doing the listening and the interpreting of the demands of nature? Is there just one way of hearing and understanding them? Dryzek extends agency (though not subjectivity) to nature: 'nature is not passive, inert and plastic. Instead, this world is truly alive and pervaded with meanings.'[65] But the meanings of nature's demands cannot be regarded as unambiguous; the communications of non-human nature must exist at the non-discursive level highlighted by Coole. This is precisely why Habermas wants to exclude it from communicative rationality. If the voices of nature are interpreted in one particular and supposedly rational way, then there clearly must be a remainder – its otherness still exists, but is suppressed. If each of us has difficulty observing our own unconscious desires and motivations and lapses, how could we hope to interpret those of nature? Right from the beginning of any attempt to incorporate the non-human and non-discursive into rational discussion, there are inevitable exclusions and suppressions.

Dryzek believes that it is possible to eradicate the functioning of power in discussion: 'The only condition for authentic deliberation', he writes, 'is the requirement that communication induce reflection upon preferences in non-coercive fashion. This requirement in turn rules out domination via the exercise of power, manipulation, indoctrination, propaganda, deception, expressions of mere self-interest, threats (of the sort that characterise bargaining), and attempts to impose ideological conformity.'[66] But this requirement is impossible to meet. Those associated with 'otherness' and those who struggle to find a route into the deliberative forum or to speak when they are there, are left out from the very beginning. Despite deliberative democracy's celebration of difference and emphasis on the importance of engaging different viewpoints in meaningful discussion, it cannot incorporate many who are *already* excluded, and by denying this it misleadingly asserts that the conclusions it arrives at are assented to by all.

James Tully questions whether it is always possible to transcend our particular perspectives as communicative rationality and deliberative democracy demand. Yet, he explains, this does not mean that we are not being *reasonable*. Using Wittgenstein's work, he notes that we may regard an idea or practice or institution as reasonable, without being able to give reasons for it. It is often reasonable to take something for granted; we cannot doubt everything. And it is not a sign of irrationality that our reasons can give out: 'the exhaustion of reasons – the inability of reasons to underpin the grounds – is not in any way irrational or epistemically defective'.[67] Many of our everyday agreements are not a matter of rational justification, but a matter of sharing a way of life and a set of conventions and values. This suggests that in a diverse society it is far from easy to cut across these non-rational agreements to produce overarching rational ones.

Green deliberative democracy relies on the functioning of a rationality that is supposed to transcend our various differences. But is it possible to separate out our reasoning from our ethical conceptions of the good life? An insight of ecology, Tully points out, is that we cannot differentiate between our ideas of the good and our conceptions of justice: 'humans exist within, are dependent upon, and are members of the web of life, the innumerable ecosystems which make up the living world'.[68] Therefore our conceptions of justice are conditioned by our forms of life and our environment. This severely impedes the rational discussion that is supposed to transcend our ethical values and perspectives. For Tully, this suggests that the elevation of morality above ethics is topsy-turvy: 'Habermas has the relation between morality and ethics the wrong way round. It is ecological ethics that is global and universal whereas deontological

morality is a limited, species-centric, or ego-centric, perspective.'[69]
Tully is correct that it is the opinion of many ecologists that there is
a universal ecological ethics, but if we locate human beings within
their diverse environments and challenge the existence of one
overarching environment we could argue that there is *no* level upon
which there exists a universal understanding and upon which we can
reach agreement. According to Plumwood and Dryzek, human life
is coexistent with the natural world and this would mean that our
perspectives are suffused with our environmental positioning. But,
as I go on to discuss in Chapter 5, this would not provide us with
one common view; rather, it would splinter the perspectives with
which participants in politics apprehend and affect the world. Any
apparent consensus, therefore, is actually an assertion of a particular
perspective, a fiction that covers up the exclusions it relies upon.
The expectation of reaching an inclusive and rational agreement,
therefore, must be queried.

Democratic Dead Ends

The goal of reaching an inclusive agreement is a misleading aim
that leads democracy down a dead end. Habermas claims that the
public sphere and rational deliberation include everyone: 'the
equal respect for everyone demanded by a moral universalism
sensitive to difference ... takes the form of a non-leveling and
non-appropriating inclusion of the other in his otherness'.[70] But
the demand for rational agreement excludes otherness. Despite
deliberative democracy's celebration of difference and its professed
aim to include voices previously marginalised, it problematically
claims that all these voices can be incorporated equally into a free
discussion. The silencing of otherness and dissent through the
assertion of communicative rationality and the aim for agreement
impedes deliberative democracy's claims of inclusion.

We cannot rely on a philosopher-king to flourish a green standard.
Quite apart from the other intrinsic values of democracy, it should
be encouraged in responding to climate change. But democracy
should not base its coordinates on the aim of producing agreement.
Both eco-authoritarianism and deliberative democracy assume that

a rational agreement or shared understanding is needed to equip democracy for tackling climate change. Eco-authoritarians believe that such an agreement cannot be reached, whereas deliberative democrats believe that it can. Both are mistaken; democracy could never produce a rational shared understanding. There is no one 'rational' path to take, no overarching grand green scheme that presents the solution. Both *whether* and *how* we act to combat climate change is a difficult political choice on which there will never be full agreement. Perhaps therefore we should reconsider whether consensus is actually needed to act against climate change, and whether focusing on achieving it obstructs the possibility of effective action. As I will go on to argue, not only is democratic disagreement key to securing decisive action, but this disagreement itself is precisely what could engender engagement with the issue by larger sectors of the population.

Radical democrats argue that differences cannot and should not be overcome, and that theories of deliberative democracy head down a dangerous path by denying the permanent presence of conflict. Chantal Mouffe, for example, emphasises the ineradicable dimension of power which blocks the possibility of giving all perspectives equal access to, and manoeuvrability within, the discussion: 'every consensus exists as a temporary result of a provisional hegemony as a stabilisation of power that always entails some term of exclusion'.[71] For Mouffe, the claim that a fully inclusive consensus can be reached must be challenged: 'taking pluralism seriously requires that we give up the dream of a rational consensus which entails the fantasy that we could escape our human form of life'.[72] Any apparent inclusive agreement reached is rather a trick of power that disguises exclusion and which should be highlighted rather than disguised. The next chapter, then, considers the radical democratic approach to climate change.

5 Celebrating Disagreement
The Radical Democratic Approach

The Parallax of Climate Change

Political issues that merit particular attention from theorists are those that are not only important in their own right, but also reveal something about ourselves and our ways of living. Such issues serve as mirrors in which political institutions, subjects and norms are reflected back to us in all their glory and glitches. Climate change is such a mirror. As Mike Hulme puts it, in climate change 'we can look and see exposed both our individual selves and our collective societies'.[1] It reveals to us what other issues cannot, since it has a scale across time and space that dwarfs other items on the political agenda. What, then, do we see reflected at us as we peer into its giant complexity? What cracks does it expose in our socio-political systems? Do they buckle under its global nature and its remoteness in time?

Let us return to the situation presented by Val Plumwood, examined in Chapter 4. She points out that despite a general awareness of environmental problems, there is a failure of current governments and populations to do anything about them.[2] Anthony Giddens notes the same difficulty and suggests that electorates won't grasp the significance of climate change precisely until it is too late: 'since the dangers posed by global warming aren't tangible, immediate or visible in the course of day-to-day life, however awesome they appear, many will sit on their hands and do nothing of a concrete nature about them'.[3] He narcissistically christens this situation the 'Giddens Paradox'. If Giddens's nomenclature is slightly askew, then is he right at least in his assessment? I suggest that he may well be right, but for the wrong reasons. Giddens sees the democratic population as unable adequately to comprehend the situation, although at the same time he seems to see himself as capable of comprehending it, as he is writing about it. But rather than the inability of the demos to grasp the situation rationally and to see beyond their immediate self-interest, perhaps the problem is that the situation isn't grasped in the way it is routinely, hegemonically, construed. Perhaps the lack of decisive action should not be regarded as an exception to the way climate change is generally depicted, but rather as a symptom of it. Perhaps the problem is not the paradox but rather the *parallax* of climate change.

An astronomical term referring to the difficulties of directly observing the stars, a parallax is the apparent displacement of an observed object due to different positions from which it is observed.[4] Giddens assumes that he sees the situation as it is, and that others cannot see it in the same way. He assumes that people see the problem but cannot comprehend it correctly and, furthermore, he regards this frustrating situation as inevitable and irremediable; he cannot see beyond his own paradox. For some, a parallax indeed exists when an object can be seen in both a direct and a distorted way. But another way of understanding a parallax is of the inevitability of observing a certain object *only* indirectly, and when the different observations are *incommensurable*. For Slavoj Zizek, for example, a parallax involves a 'constantly shifting perspective between two points between which no synthesis or mediation is possible. Thus there is no rapport between the two levels, no shared space. ...'[5] Perhaps climate change is a parallax in this sense, something that cannot ever be directly observed but is seen from a multiplicity and diversity of standpoints. Giddens seems to suggest that its mirrored surface, in which we see ourselves and our politics, is itself distorted when looked at from the wrong direction, and that only some see it truly straight on, free from any distraction by its warped reflections. I contend that there is no one correct position from which to assess climate change. Perhaps this object we call climate change is a shifting mirror ball that is observed from very different positions. Its reflective surface scatters images to and from all directions; we catch only glimpses of what others might see of themselves in it.

This might suggest that disagreement about climate change is inevitable, a result of there being no one right way to see it and assess it. But should this disagreement be regarded as a problem? Theorists and policy makers of climate change often shy away from conflict, either by denying any conflict exists (ecological modernisation), or by suppressing it through the assertion of a 'common good' (green republicanism), or by explicitly dismissing democratic politics (eco-authoritarianism), or by theorising the transcendence of conflict through rational discussion and mutual understanding (deliberative democracy). The aim is to suppress conflict so that decision making about climate change can go ahead smoothly. But, as we have seen, there are problems with all of these approaches and there has been a distinct lack of conclusive decision making about climate change.

This chapter takes a different approach and argues that the attempt to reduce or suppress conflict is badly mistaken. Using particularly the work of Chantal Mouffe, I argue here that any attempt to dispel disagreement is an attempt to eradicate politics. I go on to claim that democratic disagreement can be seen not as an obstacle, but rather as crucial in underpinning a revitalised environmental politics that would secure decisive action on an issue such as climate change.

Contrary to dominant understanding, decision is underpinned not by consensus but by disagreement, for without a choice between real alternatives there can be no decision. I will explain that this disagreement cannot be overcome through discussion, as implied by deliberative democracy; differences are not left behind during the debate, but rather subsist all the way through. The aim for a climate consensus – in both science and politics – depoliticises the issue and undermines the possibility of climate change politics. The assumption and aim of consensus, in the meantime, actually encourages those disillusioned and excluded by it to adopt the only alternative, a moral and identitarian extremism. Giddens is right when he diagnoses the lack of political inclination for tackling climate change. But this, I argue, might well be precisely because of the promotion of rational consensus, not despite it. As I will go on to suggest in the next chapter, the dangers of climate change – that are not tangible enough, Giddens suggests, to elicit the responses of the demos – might well be tangible if they could be seen from a more differentiated, localised perspective. I begin here by outlining the radical democratic approach and go on to examine how it might be useful in understanding the issue of climate change.

Radical Democracy

The radical democratic approach begins from the starting point that it is impossible to transcend disagreement and eradicate conflict from the political arena. In Chantal Mouffe's nutshell, 'conflict and division are inherent to politics'.[6] Conflict is actually what politics is all about; denying this is a denial of politics. In a liberal democracy, Mouffe explains, there is always an imperfect balance between liberty and democracy; there is always a risk that democratic decision will undermine individual rights and freedoms, but there is also always the possibility that individual rights – which are always the expression of the prevailing power – can undermine the functioning of democracy. There exists, then, a tension between the two concepts, a mutual contamination, a paradox. We can never expect society to be perfect – to provide complete equality, total freedom, to be fully democratic – because this is an impossibility. It is not just that this is an inaccessible ideal, but that the concept itself is essentially self-contradictory.[7] This means that any stabilisation of liberal democracy is always only a contingent expression of the tension between human rights and popular sovereignty. The status quo can always be challenged. There are always those who are excluded and unequal. Mouffe therefore rejects a so-called 'third-way' politics that attempts to move 'beyond left and right'. Such an approach erases the possibility of challenging the status quo: it attempts to turn 'how things are at the moment' into 'how things must be'. Any apparent consensus is really a disguised expression of power.

For Jacques Rancière, too, the essence of politics is disagreement, and he despairs of the eclipse of politics by the attempt to reach consensus: 'consensus … is, in a word, the disappearance of politics'.[8] Rancière diagnoses the emergence of the model of 'consensus democracy' or 'post-democracy', which he defines as 'the paradox that in the name of democracy emphasises the consensual practice of effacing the forms of democratic action'.[9] Consensus, for Rancière, presupposes that parties to a dispute are already settled and that there is no-one and nothing 'left over'. Here 'the people' are assumed to be an always-already fully formed ensemble that is complete and settled. But this consensus always omits those left out and uncounted: 'what indeed is consensus', he asks, 'if not the presupposition of inclusion of all parties and their problems that prohibits the political subjectification of a part of those who have no part?'[10] For Rancière democratic politics is not the clash between already established parties within the public sphere, but rather is about the interruption of the normal order (what he terms 'the police') by those who were previously not heard, those who had no 'part', who were assumed not even to have a voice. Politics, as he puts it, 'makes visible what had no business being seen, and makes heard a discourse where once there was only place for noise'.[11] Rancière draws attention to those identities that start out without any recognised identity or voice, but by finding one actually *constitute* politics. It is not, as Mustafa Dikec points out, that the unaccounted for are already counted as the 'excluded', because then they would already have a part of the whole. 'Rancière's unaccounted for does not mean that there exists a hidden bunch of political subjects to turn up and disrupt the police order. Everybody is counted. The unaccounted for is at once nowhere and everywhere.'[12] The unaccounted for are those who have no place at all and thus who disrupt the status quo by redefining the whole. 'This is the democratic theme', he explains, 'making a claim to *be* the whole, not to be *included* in the whole.'[13] Politics for Rancière, then, is about making the claim '*we are the people*'.

Rancière shows that the interruption of the normal order arising from beyond the margins of what is considered the inclusive public sphere is what actually constitutes politics. Mouffe explains, however, how identities within the political realm rely on the exclusion of each other, too. And she explains how rather than destroying the normal order, politics within the normal order can be revitalised by conflict.[14] Like Rancière, Mouffe draws attention to how politics involves not the interaction between already and fully constituted parties and identities but actually the very establishment of them. She uses the term 'antagonism' to discuss political conflict.

A relation of antagonism exists between political identities, but not as a simple clash of interests; political identities in an antagonistic relation are not constituted separately from each other. Rather they depend upon each other to exist and simultaneously threaten that very existence. Antagonism, in other words, exists between 'us' and 'them'; the 'us' needs a 'them' to constitute itself as an identity, but the 'them' always threatens to destroy that identity. For Mouffe, therefore, the political realm is not a neutral space for discussion between fully formed identities with different perspectives that can be rationally reconsidered. Rather, the political realm is constituted by antagonistic relations. Any attempt to overcome or suppress conflict is therefore an attempt to eradicate the very political dimension of our societies.

The question that arises now, is how the us/them relation can be rendered less violent yet not concealed? Democracy aims not at repressing but at *recasting* the us/them relation so that individuals and groups are not constantly at war, full of hatred and intent on annihilating each other: 'antagonism can never be eliminated and it constitutes an ever-present possibility in politics. A key task of democratic politics is therefore to create the conditions that would make it less likely for such a possibility to emerge.'[15] Mouffe explains that antagonism can be transformed into 'agonism', a relation in which political opponents are regarded as legitimate adversaries, rather than enemies to be destroyed. Here antagonism is not eliminated – which would be impossible – but *sublimated*.[16] This agonistic relationship is precisely what is required by democracy; democracy doesn't demand the overcoming of the us/them antagonism but rather its recognition, so that its potential violence can be kept in check: 'acknowledging the ineradicability of the conflictual dimension in social life, far from undermining the democratic project, is the necessary condition for grasping the challenge by which democratic politics is confronted'.[17]

Mouffe therefore asserts a model of political agonism in which difference is acknowledged but does not elicit violent hatred in response. Here, adversaries are 'friendly enemies' who share an allegiance to a set of democratic or 'ethico-political values' such as freedom, equality and democracy. But this doesn't imply they transcend their differences to reach an agreement on the content of these values, since they are understood in very different ways. This is what Mouffe means by a 'conflictual consensus': 'Consensus is needed on the institutions which are constitutive of democracy. But there will always be disagreement concerning the way social justice should be implemented in these institutions.... We can agree on the importance of "liberty and democracy for all" while disagreeing sharply about their meaning and the way they should be implemented, with the different configurations of power relations that this implies.'[18] Conflictual consensus underpins Mouffe's model of agonistic democracy, in which all those who are to be included within a democratic citizenry must adhere to 'democracy', but do not agree on what democracy *means*. By asserting the existence of a common allegiance to shared values understood differently, Mouffe avoids a reliance on agreement on substantive values and yet allows that *some* sort of consensus occurs, although it is a consensus which contains no *substantive* consensus.

The radical democratic approach, underpinned by theories of agonism, both acknowledges and celebrates the disagreements that exist over climate change. It suggests that conflict cannot be expunged or transcended, but it points out that to do so would be to eradicate the political itself. It problematises the assumption and aim of rational consensus that dominates climate change discussions.

Climate Consensus?

In December 2009, the Copenhagen Climate Summit grabbed the headlines in the urgency of its goal: 'The leaders of the world must agree on their self-appointed task of saving it,' states Sam Knight in his account.[19] However, Knight points out that agreeing is no easy task: 'the UNFCCC has no voting procedure. It has never agreed how to agree, so everything is adopted by "consensus" – essentially the gut instinct of those chairing the talks – which means that a delegate can wield enormous power by not saying anything.'[20] Consensus here actually stops the discussion. It stops it before it starts; if participants say nothing, if they remove their voice and their opinion from a consensus-seeking forum, they are able to preclude decision. Yet many current depictions of climate change see it as a matter of consensus, either as a goal or horizon for which to aim or as an already existing basis from which to start. Anthony Giddens, for example, demands a cross-party consensus and 'political convergence' of the issues as a starting point for climate change politics. He writes: 'Climate change ... is not a left–right issue. There should be no more talk of "green being the new red". A cross-party framework of some kind has to be forged to develop a politics of the long term.'[21] Various writers and politics echo this demand for climate change agreement and one leading left-wing think-tank recommends that in their aim to create climate-friendly behaviour, interested parties need to present the issue as if the argument had already been won: 'the "facts" need to be treated as being so taken-for-granted that they need not be spoken'.[22]

One of the reasons that political disagreement is disavowed is that (as we saw in Chapter 1) the scientific facts are expected to reveal to us what to do, emptying politics of any disagreement or decision. For Brian Wynne, the dominant framing of climate change as entirely a matter of science and technology presents policy as a matter of scientific discovery, as if the right way to tackle climate change could be 'read off' the scientific data. He notices that the initial question for climate science – Is long term climate prediction possible? – is no longer even asked, although there is still no answer and it is not even clear how an answer could be given. For Wynne, 'scientific knowledge should be received less as predictive truth-machine and more as reality-based social and political heuristic'.[23] But the dominant policy climate, he explains, depoliticises the issue, serving to alienate ordinary people from owning it, encouraging instead political apathy. Presented in this way, 'it might be considered natural that citizens would sit back, and wait to be told what they must do, rather than go out and learn as well as take their share of responsibility for what could have been presented as a more complex, multidimensional and inherently indeterminate set of human problems, which citizens and their representatives can and should help define'.[24] Wynne explains that the scientific framing invites a polarised response; either accept or deny, a response which has resulted in 'climate wars' between committed scientists and resolute sceptics. But if science is acknowledged to be always value-laden, and therefore to issue prescriptions embedded within a particular cultural imaginary, alternative perspectives are opened up. He advocates the empowerment of alternative imaginaries and meanings of climate change.

Gert Goeminne goes even further, claiming that climate science is political – and not just in the sense that the translation from scientific research data to knowledge is always conditioned by the values and perspectives of the scientist, as Wynne notices. Goeminne claims that science is political on its own terms, since the construction of scientific objectivity itself involves exclusion. Rather than negatively understanding the restriction of scientific knowledge by values, Goeminne's argument positively construes the situated character of science by politics. 'However neutral the invocation of science may seem, one can indeed not smooth out the non-neutral and very political struggle that underlies the decision on what to be concerned about.'[25] But because this is not acknowledged, he argues, politics is left with nothing but the technocratic management of the issues: 'lost in the translation from science to policy, the concernful work of composition that goes into the construction of a matter of fact is obscured in consensual decision making, leaving policy nothing but externalities to be managed in a technocratic way'.[26]

Erik Swyngedouw calls attention to the 'consensual presentation' of the problem of climate change. He explains that the assertion of consensus evacuates disagreement from climate change discussions and expunges politics in a post-political move: 'The post-political environmental consensus … is one that is radically reactionary, one that forestalls the articulation of divergent, conflicting and alternative trajectories of future socio-environmental possibilities and of human–human and human–nature articulations and assemblages'.[27] In her analysis of UK planning policy, Aiken explains how climate change is not only depoliticised, but used to suppress dissent in various policy areas: 'references to climate change … can be used to play a dangerous role in upholding the status quo and preventing meaningful dissent'.[28] She explains how disagreement about renewable energy, for example, is not valued and encouraged but rather viewed as an obstacle to be overcome: 'opposition to renewable energy projects is not viewed as a valued expression of public concern within a democratic institution but rather as a harmful obstruction to an agreed policy goal'.[29]

Swyngedouw, Wynne and Aiken, amongst others, urge the telling of different stories and the imagining of different futures to challenge dominant depictions of climate change.[30] But the possibility of articulating different stories of climate change is foreclosed by the assumption that a consensus is needed. Will Hutton articulates the received wisdom in his column. He urges the establishment of an 'intellectual consensus' to combat the lobbying of anti-environment US gas and oil companies: 'Lobbying and political intransigence are much easier to achieve when there is no intellectual consensus.'[31] Hutton regards climate change scepticism as the problem, and argues that it has become 'the new common sense'. This is despite the fact that 'almost everybody in the scientific world accepts the evidence that the record rise in atmospheric carbon dioxide is associated with rising global temperatures'. Is this denial then an avoidable aberration? Or is it actually connected to the way in which climate change is continually framed? Perhaps the focus upon consensus about climate change does not dissuade disagreement but actually encourages the growth of passionate climate change denial.

As Rancière explains, consensus is not simply an overcoming of conflict, but actually the stifling of the conflictual nature of common life; consensus reduces politics to what he calls 'the police': 'Consensus does not mean simply the erasure of conflicts for the benefit of common interests. Consensus means erasing the contestatory, conflictual nature of the very givens of common life. It reduces political difference to police-like homogeneity.'[32] He explains that new forms of extremist movements that have arisen in liberal democracies in recent years are not the exceptions to consensus but the result of its logic. Rather than being the re-emergence or remainder of past primitive ethnic hatreds – as they are often dismissively presented – they should be understood as symptoms of the expunging of democratic differences. Mouffe agrees with this logic. The assertion of consensus precludes any possibility for any alternative, and the only options remaining for many are offered by right-wing extremist groups: 'In this increasingly one-dimensional world, in which any possibility of transformation of the relations of power has been erased, it is not surprising that right-wing populist parties are making significant inroads ... they are the only ones denouncing the "consensus at the centre".[33]

This sort of extremism, these radical democrats suggest, appears as the only alternative left to consensus. We could extend this analysis to the phenomena of anti-environmentalism. Perhaps the growth of climate change denial that Hutton points to is not an unfortunate exception that has arisen contingently from a swamp of misinformation and irrationality, but rather an outcome of the very framing to which Hutton contributes. As Goeminne points out, the emergence of climate denial can be seen as a symptom of the apparent consensus of climate science: 'climate denial constitutes such a symptomatic outburst of the political in a completely depoliticised landscape'.[34] I would add to this that it is not just within science but across society in general that climate consensus is both asserted and demanded. The incessant emphasis on consensus gives only two alternatives: agree or reject. There is no room for multiplicity of values and voices, no room for disagreement within politics, but rather only exclusion from it.

Naomi Klein describes attending the Heartland Institute's Sixth International Conference on Climate Change, which she describes as 'the premier gathering for those dedicated to denying the overwhelming scientific consensus that human activity is warming the planet'.[35] But the consensus she both describes and reproduces is an agreement that excludes and then masks that exclusion as both necessary and natural. Climate change denial is not an exception to the climate change consensus, but its symptom. The depoliticisation analysed by Swyngedouw and Wynne results in the eruption of antagonism in a heightened extremism that shifts from the political realm towards moral territory.

If the us/them relations aren't seen in political terms, Mouffe warns, then there is a danger that they come to be captured in moral terms instead. 'They' are seen as morally bad, the carrier of some 'moral disease', and therefore to be regarded as unworthy of agonistic respect: 'instead of being formulated as a political confrontation between adversaries, the we/they confrontation is visualised as a moral one between good and evil, the opponent can be perceived only as an enemy to be destroyed'.[36] Those that disagree with the dominant description of climate change become tarred with the stigma of irrationality and moral badness and their claims have been regarded as socially unacceptable.[37] Climate change deniers who reject and are rejected from the consensus are often depicted as immoral, and their exclusion can be understood, perhaps, as leading to the polarisation and extremism that Klein describes so vividly. Yet Klein, it must be noted, is clear about the political nature of the anti-climate camp and the need to regard them as political opponents. She is aware that the science is not really the issue; it is the political implications of climate change that matter to them. As she explains, 'if you ask the Heartlanders, climate change makes some kind of left-wing revolution virtually inevitable, which is precisely why they are so determined to deny its reality'.[38]

Acknowledging the fiction of consensus and recognising the climate change denier as a legitimate political opponent might include more differentiated opinions within the discussion and produce a more inclusive, insightful and interesting politics that revolves less around the simplistic consensus-denial polarisation. This discussion would not end in agreement but would be more likely, I suggest, to end in a real decision.

Environment and Conflict

The irreducibility of disagreement over climate change has been partially recognised by some green theorists, but they only go so far. Graham Smith calls attention to the inevitable conflicts that exist regarding the environment. As he notes, 'judgements about the environment pull us in different, and at times competing, directions'.[39] He regards conflict as a clash of values that are often incommensurable and incompatible. For example, a rainforest has scientific, aesthetic, economic and spiritual values that compete and cannot be weighed against each other. He asks, therefore, that we consider how political institutions may be designed with this 'value pluralism' in mind: 'Contemporary polities are ... faced with the problem of developing policies in the light of conflicts between a plurality of values, many of which relate to the relationship between humanity and non-human nature.'[40] Although he agrees that there may never be final agreement on environmental questions, he assumes that nevertheless, different perspectives and values can all be brought into a non-coercive and unconstrained discussion. Deliberative democracy therefore promises, he claims, 'a political environment within which the plurality of environmental values can be effectively and sensitively assessed and considered in decision making processes'.[41]

The irreducibility of disagreement on environmental issues has also been partly acknowledged recently by John Barry. In a paper co-authored with Geraint Ellis, he emphasises the existence of conflict over the issue of wind energy and considers how agonism might bring new insights to his green republicanism. Barry and Ellis explain that the UK Planning Act 2008 has reduced the opportunities to challenge the building of projects such as wind farms. Although they regard the idea of achieving consensus over the introduction of wind farms as naively optimistic they don't move from this position to condoning the exclusion of dissenters in the political decisions about wind farms. They believe that exclusion hinders progression on the issue since it eradicates the opportunities for social learning: 'robust debate can help social learning and can provide imaginative solutions to problems'.[42] They argue that opening up the debate would help produce more acceptable and more creative policies: 'adopting a more agonistic framing ... can potentially provide a more fruitful way of securing a more acceptable settlement, support and government arrangements for such projects'.[43] However, Barry and Ellis give no acknowledgement of the difficulties that certain groups face in even accessing the debate in the first place – it is apparently assumed that it is possible to bring all perspectives into the discussion. Moreover, the differences Barry and Ellis are prepared to welcome into the debate are only differences over *how* carbon reductions are produced by the community; they suggest 'a greater range of options for communities to choose *how* (but not *whether*) they "do their bit"'.[44] Thus, they do not budge from the fixed *end* of the discussion. The disagreement is carefully delimited and the aim is an acceptable settlement on an already decided low-carbon vision: 'it may be possible to use the conflictual engagement of such disputes to the broader advantage of delivering low carbon societies'.[45] Tellingly, they describe agonism as lying somewhere *between* consensus and antagonism: 'one could define it as being concerned (like 'consensus' approaches) in seeking agreement (or at least settlement) but doing this by using rather than suppressing antagonism'.[46] This misrepresents what agonists and radical democrats assert; for them, there is no real agreement at all, ever.

These accounts acknowledge disagreement but only in a limited sense. Assuming that all can be included in the discussion ignores those who are not even allowed a voice, who have no voice, and no 'part'. And regarding the conflict as a clash of values that will be accorded equal weight in the discussion ignores the agonistic clash of identities and the othering that occurs. Political disagreement is defined by Rancière as not merely a conflict of interests regarding a certain situation but a clash over the situation *itself*. Disagreement is not the conflict that occurs when X says white and Y says black, but rather 'the conflict between one who says white and another who also says white but does not understand the same thing by it or does not understand that the other is saying the same thing in the name of whiteness'.[47] He goes on to explain that X understands what Y is saying, but cannot *see the object* that Y is speaking about. To apply this analysis to environmental politics, then, we might say that X and Y might both say 'green' but mean different things by it, and don't understand each other's claims. There is not one way to be green; rather, there is a heterogeneity of political 'greenness'. X and Y both see climate change but not from the same place; what climate change reflects is different for each of them. Climate change is an object of a parallax of which X and Y have incommensurable perspectives.

To be clear here, I am not claiming that all differences over climate change are insurmountable, even less am I claiming that they are fixed in any way. Shifts in political position and perspective continually and inevitably occur. What I am claiming is that we should be prepared for some differences to be insurmountable, since they may actually be defined in opposition to each other. Indeed, what I am claiming is that we should welcome such clashes.

Decisions and Disagreement

Dryzek considers the problem of 'deep difference' for his model of deliberative democracy. He argues that Mouffe's position precludes decision, precisely because it rejects the aim of consensus, and he complains that 'She scorns consensus as a cover for power, but at least consensus implies that decisions get made.'[48] Instead he suggests a separation of deliberation and decision pointing to 'the desirability of loosening the connection between the deliberation and decision moments of democracy in a divided society'.[49] He explains that decision often overwhelms deliberation and he argues that democratic deliberation could take place in a public sphere that is distanced from the sovereign authority who takes decision. This would, he argues, dampen the contest over sovereign authority and make differences less fraught. As he explains, people rarely change their minds in 'hot' deliberative forums where collective decision is on the cards, and he therefore recommends that discussion occurs away from such heated debates.

However, this is completely at odds with what I claim here. Firstly, surely separating decision from debate would dampen any enthusiasm for the debate; if debate is emptied of all stakes, then the hollow that remains is flooded only with apathy and rhetoric. A hot forum, in which passions are kindled and exchanges are heated, is precisely what is envisaged as an ideal by those dismayed by the political indifference apparently revealed in low voter turnouts today. Secondly, consensus does not *underpin* decision but rather *precludes* it by inhibiting political interaction, excluding some and dissolving the differences of those who are included so that there is nothing and no one left to choose between. A democracy that acknowledges the inevitability of conflict is a political system that is more likely to hear alternative perspectives from voices that have been excluded or ignored. I suggest that such a system that cherishes, rather than resents, disagreement, is more likely to engage people with one side or the other in the discussion, and present real choice.

Decision, I assert, is underpinned not by consensus but by disagreement, for without a choice between real alternatives there can be no decision. A decision resides exactly in the disjunction between different options; this is what makes it a decision. Politics is not an 'endless conversation' with the other, but will always in the end involve a decision which, since it is a decision between incommensurable options, will always involve force and violence. Mouffe argues that because disagreement and undecidability always exist, there will always be decision.[50] I suggest that we can turn this argument the other way around and consider that for decisions to exist there must be disagreement. Without a disagreement an issue such as climate change does not even appear as an issue for debate. Disagreement cannot be overcome through discussion as deliberative democrats assert. This doesn't mean the result is deadlock, however. In a democracy the decision that is eventually (but never *finally*) made is one that not all participants *agree* with, but that they all agree *to go along* with, at least until it can be challenged and the issue re-decided. This democratic system would acknowledge that any decision excludes every option apart from one. We can never claim to have made the 'right' decision or to have come to the *only* possible conclusion. Decisions are always pervaded by the uncertainty that is necessary for their own existence: 'We can never be completely satisfied that we have made a good choice since a decision in favour of some alternative is always to the detriment of another one … undecidability continues to inhabit the decision.'[51] The fact that those who disagree with the decision nevertheless abide by it shows precisely the strength and value of a democratic system.

For a democratic decision to be legitimate, there does not need to be substantive agreement on that decision; rather, there is agreement on the importance of democracy. Mouffe rejects the view that even democratic procedures can be agreed upon. But she does acknowledge that for a democratic system to function there must be some sort of agreement; disagreement requires an interaction that surely requires, at the very least, an agreement about how to interact. This is what she locates in her 'conflictual consensus'. In sharing an allegiance to democracy, political opponents surely agree to abide by the decisions that are taken within that democracy, in so far as they are considered legitimate. This is vital because, as we have seen, there will never be full agreement on any political decision. To act to tackle climate change we need decisive action, secured by a decision that not everyone agrees with, but to which all can agree to adhere.

To make a political decision, then, it is not just that there will be disagreement, but that there must. Collective action relies on political decision and therefore needs disagreement; thus the assumption that democratic disagreement hinders political action is mistaken. Political participants should be convinced not of the importance of *overcoming* disagreement, but rather of the importance of *disagreeing*.

Green Ends or Democratic Means?

The apparent problem for environmentalists who value democracy, and democrats who are concerned about the environment, is that there is no given or predictable outcome in a democracy. Attempts to try to fix the decision all inevitably undermine their own project. As soon as deliberative democrats or green republicans assert that the agreement reached through the debate must be one that is environmentally friendly, they have pre-empted the discussion and undermined the democratic possibility. It isn't possible to predict the outcome of a democratic debate without actually having it.

Green democratic theory thus appears to flounder on itself; the options seem to be *either* a whitening out of the greenness *or* a stifling of democracy and a suturing of the decisive moment. The reason for this stifling and suturing is the concern that democratic disagreement will produce the wrong, ungreen, outcome. If green ends are to be ensured, however, then democracy becomes regarded as merely an instrument to reach them. Rather than a political mechanism giving potential voice to dissidents to challenge the status quo, democracy is reduced by green instrumentalism as a tool to hack through the thorny difficulties preventing the move towards a climate friendly place. But it is unclear what is supposed to happen to this tool once we arrive at this locale; is it dropped beside the way and left to rust?

Alex Latta describes the problem: 'On the one hand, there is a widespread conviction that green politics should be fundamentally committed to democratic process. On the other hand, evidence of rapidly advancing ecological degradation has led to an overriding concern for achieving concrete ends.'[52] Green thinking, he explains, has tended to focus upon the latter concern, and thus it is dominated by the overriding concern with achieving the concrete *ends* of environmental sustainability rather than the democratic *means*: 'democracy is valued as a mechanism for realising green ends rather than as a sphere in which a broad spectrum of citizens can engage in fulsome debate'.[53] Latta is concerned with how citizenship has tended to be understood not in terms of political participation but rather as a way of embedding environmental duties and fostering green consciousness, and thus its 'democratising impulse' is suppressed. He draws attention to the 'discordant voices and conflictive encounters' that are forgotten by green thinkers.[54]

Latta draws from the important work of Douglas Torgerson, who highlights the problematic focus on ends over means in green politics. Torgerson believes that green theory should attempt to disentangle itself as far as possible from green ideology, which can block critical thought and push aside important concerns in the pursuit of green goals. He thus questions the metaphor of a green movement, which he sees as prone to instrumentalism since the concept of movement constructs a distance between means and ends: 'the very idea of building the green movement has distinctly instrumentalist overtones, suggesting the construction of a device strategically designed to effect social change'.[55] Torgerson wants to celebrate the intrinsic value of political debate and the 'joy of performance'. He supplements the metaphor of movement with a different metaphor of place, asserting the existence of a 'green public sphere' which is not a path but rather a space in which citizens can participate in both functional and constitutive politics as well as a 'performative politics' in which politics is given intrinsic value. By envisaging space rather than movement the diversity of those motivated by green concerns is motivated and encouraged: 'As a space for discussion the green public sphere is governed by not single direction, but displays an interest in a plurality of opinions, however inconvenient and troubling they might be.'[56] There is a 'we' here, but not a unified we that treks through the thorny jungle in one harmonious movement, but a 'we that partially exists': 'the green public sphere has a necessary commitment to debate; its inclination is not simply tolerance, but a cultivation and provocation of dis-agreements that will stimulate the exchange and development of differing opinions'.[57]

In his inspiring and celebratory description of the 'green public sphere', however, Torgerson seems to expect his participants to be motivated purely by a love of politics itself. But surely what motivates political interest is passion and identification with the political issues? How then might we celebrate the existence of the diverse political space that Torgerson visualises, without assuming that it emerges simply spontaneously? If we argue that it is disagreement that constitutes politics, and it is disagreement that encourages participation and engagement, then we can explain how green politics can be revitalised. Democracy cannot ensure green ends, but it can make green decisions possible by celebrating disagreement in the political realm. These disagreements shouldn't be expected to be overcome through the move towards a rational agreement, but to enliven political debate and to underpin decision. In this way we can unite the intrinsic value of democracy as a logic that creates a forum in which existing hegemony can be challenged, with a suggestion that this forum, revitalised by disagreement, is more likely to engage people with green issues than other approaches. Green ends cannot be imposed or presupposed, but they can at least be made more likely. I agree with Graham Smith, then, that what is key is opening up the political space, although I disagree with him about what this space looks like.[58]

Political Responsibility

I have described in this chapter theories that challenge consensus seeking. Rancière draws our attention to the voices that are excluded from the discussion itself and the foreclosure of their interruption by assertions of consensus. Mouffe notes how there are conflicts between identities that already partially exist in tension with each other. These theorists don't despair of such disagreement, but acknowledge that this is what constitutes politics. A radical democratic approach, based on these theories, is a useful approach in tackling climate change. For there will always be irreconcilable disagreements in environmental politics. There are disagreements about the question of *how* climate change should most effectively be tackled: does tackling climate change require a ban on air-travel? Should geo-engineering be used regardless of possible dangerous side effects? Does a greener world have a place for nuclear energy? What priority should it have over other political issues and goals? But there are also more radical disagreements over the reality and priority of climate change: What is the meaning of 'green'? Who should be included in climate change discussions? Does 'nature' have a voice? Decisive answers to these questions are obviously required if a collectivity is to act against climate change. My argument is that these questions can be answered, but what is importantly entailed by a radical democratic framework is an acknowledgement that there will always be those who disagree with the decision. This disagreement is too often disavowed by the presentation of the issue of climate change as a matter of consensus.

In many depictions of climate change, consensus is seen as both starting point and end point; it sits strangely, as both ideal aim and already existing. Conflict is carefully contained and disagreement is only allowed to go so far. The ostensible paradox is that despite this apparent consensus there is no consensus; although there is a general awareness of the issue and science of climate change, there appears to be no real grasp by populations that climate change demands urgent attention. If, however, we see climate change as a parallax this situation becomes less puzzling. Climate change is something that is viewed from different perspectives that are sometimes incommensurable; there is no *one* direct and correct perspective that can see the issue for what it really is. If there is to be collective action on climate change, then, a decision must be taken that can never satisfy everyone. However, this does not necessarily inhibit collective action; part of the way in which democracy works is that there is an agreement to abide by the decisions taken.

Ed Davey, secretary of state for energy and climate change, recently suggested that 'it is time to take the politics out of wind power'. Yet in the same speech he also stated that 'not enough has been done to consult, to communicate, to give people a stake and a say in what happens in their area. That is starting to change.'[59]

Davey seems to be saying that it is time to take party politics out of the discussions for alternative energy and he seems genuinely to be hoping to create political engagement. But, I have argued, this means that climate change discussions need revitalising through the celebration of disagreement. We need to put the politics back into wind and to incorporate real disagreements within environmental politics. Only this way will decisions be made.

This is the opposite to what is commonly argued. The Environmental Audit Committee, for example, demands political consensus and welcomes the creation of the Committee of Climate Change since it will help 'to depoliticise the consideration of potentially necessary but controversial measures'.[60] Despite these calls for depoliticisation, climate change remains a highly political issue.[61] I suggest that acting in response to the varied and complex effects and causes of climate change demands political decisions that are situated upon conflict, and thus are never easy, straightforward and permanent. Only by ceasing to shift decisions away from politics, and by acknowledging the difficulty and lack of finality of these decisions, can committed collective action be undertaken. Mouffe suggests that it is precisely because political decisions are never straightforward, but are infused with 'undecidability' that they demand the taking of responsibility: 'bringing a deliberation to a close always results from a decision which excludes other possibilities and for which one should never refuse to bear responsibility by invoking the command of general rules or principles'.[62] It is this political responsibility that is often absent from discussions about climate change. Challenging consensus is not politically irresponsible; it is not a shrugging off of environmental and political consciousness. Challenging consensus is rather a salutary beginning in addressing climate change.

What we need, I argue, is a *real* politics of climate change, not the continual search and supposition of a consensus that precludes such politics. I suggest we can bring the environment into our politics not by aiming for agreement about its value, but by positioning the environment as a source of political disagreement and therefore an important instigator for proper political decisions. Once we turn to the environment we see that it actually plays an important role in underpinning different perspectives and identifications. Human embodiedness and situatedness affirms the myriad positions and perspectives from which the world is experienced, and the heterogeneity of meanings of categories such as 'nature' and 'climate'. In the next chapter I consider how different perspectives on climate change arise from and influence collective identifications and outlooks, and how these differentiated understandings can revitalise politics.

6 Political Identity and Climate Change
Being Green

Dystopian Depictions

For many, the term 'climate change' instantly invokes a devastating scene of a parched smoking terrain and dusty skies, of volatile weather and haphazard flooding, of pollution and war and misery. These images are embedded in our minds by interlocking narratives. Renowned scientist and futurologist James Lovelock paints a truly dystopian picture of the overheated world of tomorrow: 'The consequences for humanity could be truly horrific, if we fail to act swiftly the full impact of global warming could cull us along with vast populations of the plants and animals with whom we share Earth ... there might – in the 22nd Century – be only a remnant of humanity eking out a diminished existence in the polar regions and the few remaining oases left on a hot and arid Earth.'[1] Lovelock's description is echoed by many others. 'One problem facing humanity is now so urgent that, unless it is resolved in the next two decades, it will destroy our global civilisation: the climate crisis,' insists Tim Flannery.[2] And US politician Al Gore writes: 'Our climate crisis may at times appear to be happening slowly, but in fact it has become a true planetary emergency and we must recognise that we are facing a crisis.'[3] Environmental activists produce documentaries that depict climate change as a dystopian catastrophe just beyond the horizon.[4] And in various forms of popular culture, climate change forms the futuristic background of an environmental apocalypse.[5]

Who constitutes the 'we' in Lovelock's statement? What climate change means, and how and whether it should be addressed, depends upon where we live and what we eat and how we travel and who we vote for. Yet this more differential framing is disavowed by the way that climate change is repeatedly presented as a global catastrophe in waiting, which will render us all victims regardless of our specificities, ironing out our different forms of life into a thin matter of survival. 'Our condition, our comfort, is killing the polar bears,' states one news reporter 'and it is going to kill us too.'[6] 'We' are invoked in such framings as a global humankind for whom guilt and innocence are irrelevant, and who must put differences aside and cooperate as one unified community. As well as victim, humanity is villain and hero all rolled into one, a protagonist who, almost-already too late, comes to a realisation of the truth of climate change.

The various dystopian depictions of climate change, fictional and non-fictional, share the image of climate change as a catastrophic *thing* sitting, hulking, in humanity's future. Lovelock's sense of urgency is perhaps understandable, but his description, I believe, should be queried. For climate change is not one process with one outcome; it is not as simple as this suggests. It is not a wave, hurtling toward the planet, crashing across all its shores simultaneously. Climate change will affect populations and environments diversely; its impact will vary – already varies – enormously.[7] This is not to say that 'we' cannot and should not act to tackle climate change, but rather to suggest that combating climate change is not a matter of acting as one giant army fighting one giant enemy, but of acting in multiple specific ways against an adversary with many faces.

This book has asserted that there is no one rational path to take to tackle climate change, no grand green scheme to which everybody concurs. There will never be a mutual understanding about what climate change means and how, and even if, it should be combated. The myth of consensus on climate change stifles collective decision and action on the issue. Controversially perhaps, I have gone as far as suggesting that participants in the politics of climate change should be convinced not of the importance of *overcoming* disagreement, but rather of the importance of highlighting and *heightening* disagreement. The question considered in this chapter is how a radical democratic approach to climate change might be implemented. I explain that the frequent depiction of climate change as catastrophic tragedy, while conveying the potentially devastating impact of climate change and the urgency to address it, actually does nothing to provoke political decision and collective action on any level. On the contrary, as Erik Swyngedouw points out, these depictions have depoliticised the issue and removed it from democratic politics. I argue that we should challenge the simple portrayal of climate change as one tragic catastrophe facing humanity and see it instead as a diversity and multiplicity of issues that are understood in many diverging, overlapping and conflicting ways. By acknowledging these differences and the inevitable disagreements stemming from them we could, perhaps, revitalise climate change politics.

In the next section I will examine how nature and climate are *political* categories, and that acknowledging their political construction reveals their heterogeneity. I go on to show how the tendency to flatten out these differences in one unified and simplified account excludes a variety of perspectives, and that if there is to be a politics of climate change, any overarching narrative needs to be supplemented by a differentiated account, one in which each audience sees something unique, one in which the audience themselves are part of the performance. By paying attention to the meaning of nature and climate for our manifold collective identities we can bring our environment into our political discussions. Collective identifications serve as important cauldrons of green thinking and actions. By being proud of our identities and differences but respectful of others we can nurture green politics. I conclude that attention should be deflected away from global agreements and towards *local identities*. Democratic disagreement between agonistically opposed identities may be able to underpin real collective decisions.

Political Nature

What is distinctive about green thought, we are told, is that it puts 'nature' at the centre of attention.[8] Nature, as Val Plumwood points out, is not a descriptive but a political category. The term 'nature' is, in western culture, often understood in contrast to 'culture' and 'reason'; it is reason's other. It is everything that reason is not supposed to be. In the dualisms that pervade western culture (public-private, culture-nature, mind-body, man-woman), one term is always 'other' to the fully positive one; the category that is 'other' is what is expelled by the primary category in order to define itself. As Elizabeth Grosz explains, 'dichotomous thinking necessarily hierarchises and ranks the two polarised terms so that one becomes the privileged term and the other its suppressed, subordinated, negative counterpart'.[9] Thus, for Plumwood, nature is 'a sphere formed from the multiple exclusions of the protagonist-superhero of the western psyche, reason, whose adventures and encounters form the stuff of western intellectual history'.[10] Cindy Katz and Andrew Kirby also highlight the problematic othering of nature that is accompanied by its mythologisation: 'Nature is another "other" at once both wild and pure, both fascinating and frustrating in its construction as that which resists intervention.'[11] It is this othering of nature, these theorists argue, that has led to the environmental crisis. Western capitalist societies have regarded nature as something that can be, and should be, conquered and colonised in the name of reason, a passive and lowly sphere that reason wants to leave behind and, at the same time, to master. 'By "othering" nature in science and art and society, we participate in the ideological practice that enables us to plunder it.'[12] Nature, along with other categories in dualist opposition to reason, is oppressed in a pattern of human domination over both human and non-human groupings.

Nature, seen in this light, is not the given, passive, lumpen object it is cast (out) as, but rather can be seen as a political category that is constructed within particular socio-cultural frameworks. Nature is constructed by culture physically, through human processes such as agriculture and urbanisation and exploration. Ideas of nature are also constructed by our cultural frameworks. This works the other way too; our cultures are conditioned by their location within our culturally constructed nature: 'our conceptions of the aesthetic attractions and value of the natural world have themselves been shaped in the course of our interaction with it'.[13] Nature and culture are not two separate arenas; they are symbiotically related. Kate Soper poses the question 'What is Nature?' and explains that this question may seem absurd, since nature is something that is at the same time both fully familiar and entirely elusive and seems to defy any attempts at definition.[14] She explains that while 'nature' is a construct of human culture, this is not to deny that there are entities and processes that exist independently of human discourse: 'while we might agree that everything by which we are surrounded (everything "touchable") is "cultural construction" ... it is equally important to acknowledge nature in the realist sense of causal powers and processes enabling and limiting cultural work'.[15] Soper, therefore, is a realist about nature but also a humanist who seeks to expose the oppressive naturalisations of power relations. She points out that there are no normative implications in nature itself; it cannot in itself provide us with environmental policies: 'it is a mistake to view ecological politics as a matter of having the "right" attitudes to the "otherness" of nature'.[16] Ecologists who intend to 'read off' the value of the environment from the material reality of nature are naively assuming its meaning to be given rather than culturally mediated.

Once we note the culturally conditioned meaning of nature we see that there will inevitably be a plurality of perceptions concerning its value and meaning. 'Nature' is not a given and fixed signification but rather has a heterogeneous meaning. As I have already noted in passing, it is a mistake to believe that a heightened consciousness of the inextricable human embeddedness within non-human nature would produce a human community with *one* unifying ethical-environmental perspective. Connecting humans with their environment would not result in one common view; instead it would splinter the outlooks through which the world is apprehended and affected. Human populations are never homogeneous and unified entities, but rather encompass a myriad of intersecting identities of ethnicity, religion, class and gender. These differences arise in part, I suggest, from the various particular environments in which members reside. 'Nature' is not one unified phenomenon but should rather be understood as being comprised of a multiplicity and diversity of *environments* which have a heterogeneous impact upon meanings, perspectives and values. What is 'good for the environment' thus varies across space and time, culture and identity. As Iris Young states, 'Every human existence is defined by its situation.'[17] Young considers the historical, cultural, social and economic aspects of situations, but we could add to this list the natural aspect of the situations in which humans live. Environment is entwined with human subjectivity and the particularities of human experience are embedded in their material reality.

As Gabrielson and Parady note in their work on 'corporeal citizenship', attending to the embodiedness of citizens shows that differences are inevitable; we are each uniquely situated in diverse environments that partly comprise our plural perspectives. 'Corporeal citizenship recognises the centrality of the environment to human subjectivity by acknowledging the variety of places that bodies inhabit and the diversity of human relations with the natural world.'[18] Gabrielson and Parady suggest that a recognition of human corporeality highlights the important link between citizens and their environment, and in this way the very concept of citizenship is inevitably 'greened'. But what is important for my argument here is their observation that human situatedness within bodies and material realities proliferates perspectives and meanings. A multiplicity of environments produces a diversity of perspectives, values and identifications. Our perception of our environment is bound up with our understanding of who we are, and who we understand ourselves to be affects our perceptions of our environments. In this way, the environment becomes an important ground of difference and disagreement.

Climate, too, as Mike Hulme shows, is socially constructed. The idea of climate is entirely abstract; it cannot be directly experienced, like rain or wind, and it cannot be measured like temperature. It is entwined with culture and the meaning of climate varies across human societies. Hulme explains that we cannot reduce it to a meteorological measurement, since climate 'is an idea that carries a much richer tradition of meaning than is captured by the unimaginative convention that defines climate as being "the average course or condition of the weather"'.[19] Climate can be something to be mastered or something to be protected. And climate change might be seen as a threat or as having creative potential: 'the ways in which the story of climate change and human civilisation has been told have ... changed over time. The dominant trope in this story has been one of climate change as threat, and yet dissenting voices have emerged which emphasise the creative potential for societies that can be found through changes in climate.'[20]

Bringing nature into our awareness does not therefore augur a unity of thought. De-othering nature would not bring us to the same recognition of some deep truth but would multiply and diversify perspectives. As some green thinkers are aware, there are a plurality of ways to be an environmentalist.[21] As John Dryzek notes, there has been a 'proliferation of perspectives on environmental problems'.[22] This is sometimes not appreciated.

In his article 'Confessions of a recovering environmentalist', Paul Kingsnorth explains, with Wordsworthian eloquence and sentiment, his powerful emotional relationship with nature. He writes: 'the downs, the woods, the rainforest, the great oceans and, perhaps most of all, the silent isolation of the moors and mountains ... took hold of me somewhere unexamined'.[23] Kingsnorth became an environmentalist because of his deeply held desire to protect these wild places. But the current environmental movement, he complains, does not have any place for such an attitude. Its total focus upon climate change and carbon emissions, and its unrelenting drive to reach sustainability, will involve plundering the earth's resources still further and damaging its natural landscapes and its untouched wilderness through the erection of gigantic wind farms and acres of photovoltaic panels. Kingsnorth nostalgically yearns for the environmentalism of yesteryear which was 'genuinely environmental'; the goal of this real environmentalism was 'saving nature from people' and the members 'were, like me, ecocentric: they didn't see "the environment" as something "out there", separate from people, to be utilised or destroyed or protected according to human whim. They saw themselves as part of it, within it, of it.'[24]

This attitude is obviously genuine and passionate, and one imagines it chimes with many other eco-centric, 'dark' or 'deep' greens. Yet it is not the correct and only perspective for those concerned about the environment. For a start, as Soper explains, this eco-centric perspective ignores the continual and pervasive physical reach of human society into the supposed wilderness. And, secondly, it ignores the socially conditioned and culturally located meaning that both Kingsnorth and his environment are entrenched within.

Kingsnorth implies that those who don't feel the way he does are either blindly irrational or selfishly immoral. But '*the* environment' is not a given thing with a fixed meaning and value. There are multiple *environments* that both construct and arise from human cultures.

How can we combine recognition of such passionate identifications as Kingsnorth describes with an awareness of the legitimacy of other perspectives? I suggest that radical democracy, as outlined in Chapter 5, helps us to do just this. Radical democracy celebrates disagreement between opposed identities, but attempts to ensure that this disagreement doesn't erupt into violence and aggression. The 'other' in a radical democracy is not an enemy to be destroyed but a legitimate adversary. Such an approach demands that we contest dominant overarching narratives of climate change with localised and differentiated ones. De-othering nature, recognising it as interlinked with culture, would interrupt its presentation as one homogeneous thing and enable political contestation over its conventional portrayal.[25] However, as I go on to examine next, depictions of climate change as catastrophe and tragedy iron out differences, constructing nature as an externalised and passive thing, forgetting our myriad specificities.

Apocalypse Soon

The Age of Stupid, the film directed by Franny Armstrong, was watched by a wide audience that was out of proportion to its small budget. The film is set in a climate-ravaged and environmentally depleted world of the future. Pete Postlethwaite plays the part of an archivist based in an ice-free Arctic who is looking back upon footage leading up to 2008 and asking why humanity didn't stop climate change when there was still the chance: 'We could have saved ourselves, but we didn't,' he sighs. Although the implicit message here is that it is very nearly too late to act against climate change, in this scenario it is actually *already* too late; the catastrophe has occurred and all the audience can do is feel sorry for its victims and the lonely archivist. Such a story is not so unusual in narratives of climate change, which tend to interweave elements of emergency, catastrophe and tragedy into one overarching homogenising account.

For Craig Calhoun, *emergency* has become the prevalent term in understanding the various types of conflicts and catastrophes that cause human suffering. He argues that this term naturalises what are actually products of human action. An emergency is something sudden, unpredictable, short-term and abnormal although, as Calhoun explains, in fact what the term is used to refer to are clusters of events that have developed gradually and predictably over substantial periods of time. The term also separates the events from their causes, disguising the chain that links entirely preventable causes to their catastrophic results. This dislocation renders the emergency as an unforeseeable assault of nature, an unfortunate exception to our normally ordered world. For Calhoun, emergencies have arisen as increasingly important features in dominant descriptions of the world: 'international and global affairs have come to be constructed largely in terms of the opposition between more or less predictable systems of relationships and flows and the putatively unpredictable eruptions of emergencies. This reflects both the idea that it is possible and desirable to "manage" global affairs, and the idea that many, if not all, of the conflicts and crises that challenge global order are the result of exceptions to it.'[26] He explains that depicting events as emergencies legitimises responding to them through intervention: 'emergency is a way of grasping problematic events, a way of imagining them that emphasises their apparent unpredictability, abnormality and brevity and that carries the corollary that response – intervention – is necessary'.[27]

Calhoun is referring primarily to famines and wars, but I think we can tentatively apply his analysis to climate change too. Climate change is portrayed, I suggest, as a *single* event, rather than a cluster of different phenomena and issues, and as an *abnormality* of both the planet earth's existence and our own, an exception that demands that we put our normal order on hold for a while, but don't actually seek to transform it. Although there is a general awareness that climate change is a product of human actions, by envisaging it as an emergency, the links between actual actions today and the process of climate change are severed. Contrary to Calhoun's examples, however, climate change is understood as an emergency that has not yet happened, a ghastly spectre of the future that demands intervention today. The call goes up for drastic measures and because these are not forthcoming, disillusion sets in. 'Eleven days to save the planet,' proclaimed *Prospect* magazine in the build-up to the UN climate summit in Copenhagen. But the planet wasn't saved and, after the disappointment, climate change has slipped off the front pages. By disavowing its differential connection to our current ways of life the call to a response to climate change is answered not by realistic and difficult policy decisions but by panic and despair over the tragedy-to-be and then an instant forgetting. Portraying climate change as an emergency seems to sever the possibility of its containment and make its devastating consequences all the more likely.

Another way in which depictions of climate change overstretch Calhoun's concept of emergency is that they frequently tell the story of a widespread event, one that isn't just *one* event but *the* event that affects the whole of humanity. In these stories, climate change is not just an emergency but a *catastrophe*, a sudden and violent event that shakes the entire planet. As Paul Kingsnorth notes, there seem to be only two futures possible: Saving the World or Apocalypse Now.[28] Al Gore, for example, writes that the climate crisis 'is a true planetary emergency ... unless we act boldly and quickly to deal with the underlying causes of global warming, our world will undergo a string of terrible catastrophes'.[29] In portrayals such as this, fear and urgency are elements that combine to homogenise both nature on the one hand and humanity on the other. Erik Swyngedouw provides a list of newspaper headlines proclaiming the advent of environmental catastrophe and draws attention to the 'apocalyptic imaginaries' that pervade narratives of climate change. These imaginaries invoke the vulnerability of nature and homogenise the whole of humanity: 'Irrespective of the particular views of Nature held by different individuals and social groups, consensus has emerged over the seriousness of the environmental condition and the precariousness of our socio-ecological balance.'[30] He explains that by repeatedly portraying climate change as a universal humanitarian threat, we *all* become victims. Swyngedouw notes that the IPCC is, to be sure, careful to point out that the poor will be hit foremost and hardest, but merely goes on to urge the tackling of climate change in the name of the poor. The homogenising depiction of climate change negates the social heterogeneity of the world's population and the dissimilarity of the impacts and perspectives of climate change. It 'cuts across the idiosyncrasies of different, heterogeneously constituted, differentially acting, and often antagonistic human and non-human "natures"'.[31] Brian Wynne agrees that the globalising and homogenising depiction of climate change disguises the fact that climate change, for some people in some places, is already happening. 'For many people,' he notes 'apocalypse has indeed already arrived.'[32]

Philip Hammond and Hugh Ortega Breton point out in their analysis of films about climate change how the fear of catastrophe is used today in an attempt to engage populations. As they note, however, this tactic is not very successful: 'the use of fear of catastrophe appears to be ineffective because the severity of predicted dangers does not correlate with everyday knowledge and experience of present circumstance'.[33] In other words, the more catastrophic the depiction, the less real it is and the less engaging it becomes. They argue that filmic depictions of climate change also frequently feature the experience of personal loss to provoke a reaction from the viewer. This may not sound so unusual in a disaster movie such as *The Day After Tomorrow*, but Hammond and Ortega Breton are also referring to documentaries such as *The Age of Stupid* and Al Gore's *An Inconvenient Truth*. Their observation is that all these films invoke personal sadness and nostalgia for a lost past, but that this invocation avoids any requirement for genuine political argument: 'In all three films, climate change is pulled into an affective human-interest frame via association with personal loss ... the question of political agency in the present is implicitly rendered problematic.'[34] Perhaps emotional appeals are seen as more important that information and rational argument, Hammond and Ortega Breton suggest, because political actors today have difficulty in making overtly political arguments.

In *The Age of Stupid*, the archivist plays footage of our present ways of life, seen from the perspective of future climate change catastrophe. Climate change is constructed here as a tragedy-to-be, the loss of the ways of life that we are leading now. We become nostalgic for our own present. The theme of *tragedy* recurs in accounts of climate change. For Stephen Gardiner, for example, 'the global warming problem has a tragic structure'.[35] As Aristotle told us, tragedy arouses feelings of pity and fear in its audience, and then purges them in a process of emotional healing known as *katharsis*. On one interpretation, Aristotle suggests that katharisis can help those people who feel an excess of fear or pity to find balance.[36] The portrayal of tragedy allows its audience to purge themselves of inappropriate levels of emotions so that they won't indulge in them in their ordinary lives. Do certain depictions of climate change hope for a similar cathartic outcome? The tragic portrayal of climate change seems to be about ourselves rather than the planet. Tragic portrayals hold value for our own emotional balance but do nothing to convince us to change the story line.

Audience Participation

Douglas Torgerson notes the dominance of tragic portrayals in environmentalism. 'Arising in an atmosphere of crisis, green politics seems to fit the story line of tragedy', the human domination of nature has tragic consequences, and 'a remorseless destiny unfolds despite heroic action'.[37] But alongside the tragic depictions of climate change, Torgerson counterposes a different dramatic mode. He argues that *comedy* may be a useful strategy in addressing environmental issues. Most green thinkers, of course, find the issues they are concerned about to be no laughing matter and a comic depiction of climate change seems to have a problematical lack of appropriate seriousness. But the comic mode, Torgerson explains, mocks human pretension to heroism and grandeur, the notion that humans are somehow special, and it tackles the human colonisation of nature: 'In contrast to the tragic portrayal of humanity in terms of godlike aspirations, comedy accentuates human faults and limitations. The focus shifts from the transcendent to the finite; foolish notions, ridiculous situations, the less exalted functions of the body, the inexhaustible range of human foibles.'[38] An insight of comedy is the recognition of paradoxes that cannot be rationally solved but might be creatively bypassed: 'Politics ... is a comic juggling act that defies fixed concepts.'[39] Torgerson draws attention to the comic moments and gestures that exist already in green politics, the absurd gestures of some environmentalist groups and the carnivalesque atmosphere of some protests. While tragedy culminates and hinges upon its ending, comedy has no such finality; the end of a comedy is where the comedy finishes; it exists within the joke and the laughter itself: 'comedy has no final resting place, but seeks always to renew comic tension, to keep the story alive'.[40]

This comic portrayal fits Torgerson's call for recognition of the *intrinsic* value of democratic politics rather than the prioritising of its instrumental value in reaching a goal. Enjoying the jokes would foster an enjoyment in the performance of green politics itself. But because comedy has no ending, there is no place for a moment of decision. We need an ending in politics, however impermanent. The problem with both genres – of tragedy and comedy – is that the story remains the same wherever it is performed. There is the same set of characters, the same hero and villain. Although Torgerson's comedy seems to hint at a less simplified account, I think we need to move beyond single narratives towards a recognition of the multiple overlapping threads and scraps of stories about climate change. Climate change tells a different story for each of its audiences. Each time it appears on stage it is unique, with a different beginning and ending. It reacts to its audience, who become part of the story itself, and to the situation in which it is set. Climate change is not tragedy or comedy, or a simple combination of the two, it is an *improvisation*, at times tragic and at other times amusing, for which there is no one definitive meaning and no one cast of actors.

As Swyngedouw concludes, the frequent warnings of dire ecological apocalypse knit together to preclude any genuine political framing of climate change: 'the presentation of climate change as a global humanitarian cause produces a thoroughly depoliticised imaginary, one that does not revolve around choosing one trajectory rather than another, one that is not articulated with specific political programs or socio-ecological project or revolutions'.[41] The depoliticisation of climate change, Swyngedouw explains, means that decision making is left up to experts and elites and thus the status quo is allowed to go unchallenged, plucking the issue away from democratic discussion and those with different ideas: 'It does not invite a transformation of the existing socio-ecological order but calls on the elites to undertake action such that nothing really has to change, so that life can go on as before.'[42] He adds that, in this way, 'Politics becomes something one can do without making decisions that divide and separate.'[43]

Climate change is often conceived as one global issue, but its impact varies greatly at the local level. Climate change is not one single problem. For some, climate change means flooding, for others drought, for some mitigation and others adaptation. We are affected by it in different ways but we also contribute to it differently.[44] As Mike Hulme puts it, 'the idea of climate change means different things to different people in different contexts, places and networks'.[45] In his well-cited book, Hulme counsels against regarding climate change as simply a material 'fact', regarding it instead as a 'mutating idea'. For Hulme, climate change has a socially constructed meaning, thus there is no one vantage point from which it can be accurately accessed: 'Our discordant conversations about climate change reveal … all that makes for diversity, creativity and conflict within the human story ….'[46] He explains that by constructing climate change as the 'mother of all problems', by seeing it as *the* problem, we have set ourselves an unassailable task: 'We have created a political log-jam of gigantic proportions.'[47] Yet Hulme is not pessimistic about the future; he suggests that the idea of climate change has the potential to teach us things about ourselves. It is not a problem that has a solution, it should rather be seen as a potentially helpful *idea*: 'the idea of climate change should be seen as an intellectual resource around which our collective and personal identities and projects can form and take shape'.[48] If we continue to try to master the climate system we will inevitably always be frustrated: 'our engagement with climate change and the disagreements that it spawns should always be a form of enlightenment'.[49] By focusing upon the creative potential of climate change, however, Hulme seems to undermine his own claim that climate change is a matter of disagreement. For some, surely, will not see the creative potential but only the destructive capacities of climate change.

How might diverse and overlapping perspectives transform

responses to the problem of climate change? Swyngedouw and Wynne and Hulme suggest opening up climate change politics to different stories and perspectives: 'a new radical politics must revolve around the construction of great new fictions that create real possibilities for constructing different socio-environmental futures'.[50] Instead of the tragic and catastrophic depictions that show climate change as one thing impacting us uniformly, perhaps we need films showing the different impacts of climate change in different regions. Instead of urging politicians to come to international agreements, perhaps as citizens we need to consider how we ourselves are situated in our particular environments and what climate change means to our particular self-understandings and views of the world. This might help an engagement with climate change in a more politically differentiated way. Rather than focusing on global agreements, then, we need to shift our attention to collective disagreements.

Political Identifications

Climate change discourse seems to function primarily at the global level of humanity and construct 'us' as humanity. George Monbiot argued that at the UN climate summit in Copenhagen 'humankind decides what it is and what it will become.... This is about much more than climate change. This is about us.'[51] But nature and culture intersect in particular ways with different individuals and groups and help construct different identifications. People will have very different understandings, concerns and hopes about climate change. What 'we' will suffer and what 'we' should do depends upon who 'we' are.

Naomi Klein illustrates the importance of political identification in her article on climate change denial and politics. She explains that Americans used to care very little about the issue of climate change. But now there are a group of Republicans that care a great deal about this issue, albeit in a negative sense; they are passionately in *opposition* to it and are dedicated to exposing it as a Trojan horse for socialist ideals: 'there is a significant cohort of Republicans who care passionately, even obsessively, about climate change – though what they care about is exposing it as a "hoax" being perpetrated by liberals to force them to change their light bulbs, live in Soviet-style tenements and surrender their SUVs. For these right-wingers, opposition to climate change has become as central to their worldview as low taxes, gun ownership and opposition to abortion.'[52] These right-wingers identify with climate change denial. Their identification is not a matter of rejecting scientific fact, but of regarding it as an issue that is key to their collective self. As she points out, if an issue becomes central to someone's *identity*, then rational argument has little clout in changing their stance.

Identifications are important to individuals; they provide a sense of belonging and meaning in the world and allow them to act as part of a group. Collective identifications are therefore central to politics. By uniting as a group, people are able to challenge the status quo. Identities, therefore, involve sharing something with others. However, identifications are also always precarious, vulnerable and threatened. Although identity is sameness, it is also difference. As William Connelly explains, to identify is to differentiate, and he points to 'the constitutive role of difference in identity itself'.[53] Identity only exists in relation to that which is different from it. My identity is specified in relation to things I am not. For us to be us, we need to expel what is not us: 'Identity requires difference in order to be and it converts difference into otherness in order to secure its own self-certainty.'[54] Chantal Mouffe points out how collective identities arise through a relation of antagonism in which one relies upon the other and yet simultaneously rejects them. Identities expunge what they are not, and in doing so define themselves. This relationship is antagonism, which is constitutive of identity. Antagonism is not an opposition between two fully constituted objects A and B. Rather, antagonism involves the contamination and mutual dependence of two identities. Antagonism exists where A is not fully A because of B: 'Antagonism constitutes the limits of every objectivity ... a relation wherein the limits of every objectivity are shown.'[55] We blame the other for blocking our identity, for stopping us being fully 'who we are', although this blame is illusory. There is an ever-constant threat that antagonism can turn into violence, which is why Mouffe asserts agonism as a way of transforming antagonism into a relationship that is less aggressive. The precariousness of identity reveals that identities have no fixed meaning. Their meaning arises through the continual (re)negotiation of the social structure, individual self, and environment. Nature and climate intersect symbiotically with different identities and therefore the meaning of nature and climate change alter along with our shifting self-understanding. Yet this contingency doesn't mean that identities are not important and valuable.

My proposal is that rather than seeing antagonism between identities as a problem to be expunged from climate change debate, we should notice that the disagreement that it engenders can, when properly contained by democracy, revitalise politics. By noticing how climate and nature intersect with 'us' in particular ways we can celebrate disagreement and put it to work for us. For some collective identifications climate change is a threat, and by voicing their concerns collectively they can draw attention to them. For others, climate change may be less of a priority or contain new possibilities. What 'we' should do depends on who 'we' are.

Klein argues that the left should learn from the right; the right-wing climate deniers, she claims, are quite correct in diagnosing a link between climate change and left-wing politics, and she advocates a green-left world outlook: 'Just as climate change denialism has become a core identity issue on the right, utterly entwined with defending current systems of power and wealth, the scientific reality of climate change must, for progressives, occupy a central place in a coherent narrative about the perils of unrestrained greed and the need for real alternatives.'[56] Climate change is not just one more article on the laundry list of causes for the left but rather has the potential to change society in a progressive way. Klein claims that although climate change needs strong government it also needs devolution of political control, and is therefore an issue that can serve as a catalyst for profound social and ecological change. But this means that those on the left need to integrate it into their world view. Her suggestion is that the left incorporate climate change as a central issue for their collective projects.

Can we see greenness itself as a political identity? Giddens suggest that Green politics went wrong in the 1970s and 1980s by presenting itself in opposition to mainstream parties.[57] But in presenting an alternative to the mainstream, by identifying themselves differently, I contend that this is precisely where Greens went right. As a political identity, greenness has an important place in climate change politics. It cannot get everyone to identify identically, but it should not want to. It should be pleased to disagree. Green radical democrats do not demand that everyone agrees with them, and do not dampen their views to fit with another's, but strongly and passionately argue for the policies that they believe are socially just and ecologically right. Greens are constituted as greens only by being differentiated from those who are not greens, and their identity therefore relies upon this differentiation as well as being threatened by it. If there weren't those who weren't green there wouldn't be anyone who was. As Richard Rorty notes, our solidarity is strongest where we are not everyone: 'our sense of solidarity is strongest when those with whom solidarity is expressed are thought of as "one of us" where "us" means something smaller and more local than the human race'.[58] Rather than expecting everyone to eventually agree with them, to see reason, perhaps greens and climate change activists need to recognise themselves as a strong political identity, an 'us' that cannot incorporate humanity because then 'us' wouldn't mean anything. By ejecting consensus and encouraging strong identifications with a green outlook, could we put the politics back into climate change? Of course there would be disagreement between different ways of being green. There would not be one unified environmentalist movement, although links could be formed between different groups.

It is crucial to frame climate change as a *political* issue rather than a *moral* question. For in politics we can see opponents as legitimate adversaries rather than enemies to be destroyed. In this way passions about the environment and the climate are mobilised, they spark and sustain debate, they help to foment democracy and to demand a decision on what they regard as the relevant issues.

Global Agreements, Local Identifications

How might action against climate change be motivated and then sustained over the long term? Many assume that the only possible route is via a global agreement on climate change. A report by the World Resources Institute states that 'Only through global coordination will we be able to move our society to a more sustainable future.'[59] Yet it has been repeatedly shown that such agreement is not forthcoming. The outcome of the UN climate summit in Copenhagen was at best 'ambiguous'[60] and at worst 'a failure whose magnitude exceeded our worst fears and the resulting Copenhagen Accord was a desperate attempt to mask that failure'.[61] In response to the recent Rio Earth Summit Will Hutton wrote: 'There were plenty of warm words and reaffirmations of intent – but nothing that might address the intense pressure on the natural environment.'[62] There have been different reasons given for the difficulty of securing legally binding global agreements between states. Some theorists point to problems of leadership,[63] others to the structural faults in the UN negotiations or to domestic constraints.[64] Others blame the problems of enforcement.[65] Perhaps it is time to move away from the focus upon global *agreements* and consider looking instead at 'deals' or 'decisions' that are not based upon consensus but rather acknowledge the existence of disagreement. Such deals might be more effective if they were supported by domestic publics who through a healthy political sphere had made their voices heard. A narrow focus solely on the global level can therefore also be queried.

In her discussions of climate change and democracy with reference to the specificity of the region of the South Pacific, Bronwyn Hayward agrees that climate change is not usefully understood solely on the global level, since 'climate change is not a uniform global phenomenon; it is a multiscale spatial and temporal issue.... Local communities will experience [the] impacts in differing ways and at different times.'[66] However, as she observes, the climate change discussion is firmly stuck at the global level: 'The prevailing wisdom', she notes, 'is to address climate change issues not via local debate, but through international collaborative efforts and frameworks that recognise the international causes and consequences of climate change.'[67] Hayward makes the important point that different places and peoples suffer in different ways, and that stresses on the climate will impact variously upon differentiated patterns of oppression. She emphasises the importance of *local* conversations since only communities themselves are able to consider how they will suffer: 'Local conversations about the weather are required to establish what is valued, what is vulnerable to climate change and how communities might respond effectively.'[68] These conversations, she says, can be helpfully linked to each other and to transnational discussions, but should not be conflated together.

Hayward asserts a deliberative democracy in which all voices are heard, but she recognises that power relations structure discussions and that deliberation on climate change has often been reduced to a neo-liberal mechanism to allow stakeholders to protect their private property. I suggest that we add disagreement into her account. The discussions that Hayward recommends could be revitalised by democratic conflict. There would be no *agreement* at the end, no revelation of the rational solution, but there would be a decision that would instigate some sort of collective action that possibly would take a green form. As she explains, local conversations could be linked to conversations at other levels, forming new allegiances and conflicts. These conversations may teach us something but they may heighten disagreements or create new ones.

It is surely now time to note that there is not one perspective on climate change and that there never will be. Any global agreement on such a complex issue is unlikely to be concocted, and if it is, its content would be watered down so much it would mean very little. I argue, however, that turning attention to the *heterogeneity* of perspectives on climate change could be a salutary move. And by being proud of our own identities and respectful of others, we could move towards a decision, a decision that will never be *final* but one that might be able to tackle climate change, whatever it is decided this means.

Conclusion
Beyond the Not-Plastic

The slogan emblazoned on the not-plastic bags seemed, for a while, to crop up everywhere. I am not, it proclaimed, un-green. Not-plasticness became a value so omnipresent that to be not not-plastic seemed somehow to be radical. Not-plasticness became so famous that it detached itself from the green values that spawned it, and went floating off downstream happily forgetting its environmentally friendly roots, presumably winding its way to the rubbish dump where all other consumer perishables eventually end up. Not-plastic bags are an example of atomistic approaches that engage us as consumers rather than citizens. Here, we can both sustain our cake and consume it; here, the cake is not-plastic and thus we can feel we are contributing without changing our current socio-political systems or looking through the green wash they are frequently daubed with, to consider what climate change actually means.

We can't say that not using plastic bags isn't good for the environment. But what might be a more effective strategy in tackling climate change? I have considered, in this book, the techno-economic, the ethical-individual, the green republican, the deliberative democratic and the radical democratic approaches to climate change. All, I have tried to show, provide important contributions and insights. But whereas the others ultimately spurn disagreement, radical democracy celebrates it, and thus provides what I believe is a more democratic and a more valuable approach for tackling climate change. By acknowledging and encouraging disagreement, radical democracy revitalises politics and foments engagement with the issues and the political discussion itself. The decisions that arise at precisely the points where disagreement exists can never claim to be final and complete, yet they are nevertheless the decisions that are needed to secure collective action. These decisions can always be challenged, and they are never guaranteed to be green. The outcome may be plastic, or even not-not-plastic. The temptation therefore must exist to put green ends before means and assure that climate-friendly policies are implemented. But 'climate-friendly' has no one meaning that can be discovered through rational thinking and communication. What it means to combat climate change depends upon who you are, where you are from, and where you would like to go. I have argued that if we want to improve the chances of climate change rising up the political agenda we cannot demand that people 'see reason'. But we can acknowledge the environment in our perspectives and identifications and distinguish our collective identities in distinct ways. A green radical democratic would defend their beliefs passionately but nevertheless would acknowledge another's right to disagree. By celebrating identifications in the political realm and taking note of how they intersect with nature and climate we can reinvigorate politics through difference and disagreement. Who are we, who do we want to be, and in what ways are we connected to our environments? By asking these questions about ourselves, perhaps we can rethink climate change; perhaps we can reconsider its implications, its causes and its possibilities. By noticing the disagreements spawned by changing climates in a changing world, we may be able to negotiate decisions without waiting for an illusionary consensus.

Notes

Introduction

1 'Climate change is simply the greatest collective challenge we face as a human family.' Ban Ki-moon, UN Secretary-General, remarks at the 39th plenary assembly of the World Federation of UN Associations, 10 August 2009.

2 Hulme, Mike (2009) *Why We Disagree about Climate Change*. Cambridge: Cambridge University Press.

Chapter 1

1 Plato (1961) *The Apology* in Hamilton, Edith and Cairns, Huntington (eds), *The Collected Dialogues of Plato*. Princeton, NJ: Princeton University Press, 23b.

2 Oreskes, Naomi and Erik M. Conway (2010) *Merchants of Doubt*. London: Bloomsbury.

3 Arrhenius, Svante (1896) 'On the influence of carbonic acid in the air upon the temperature of the ground', *Philosophical Magazine and Journal of Science* 41: 237–76.

4 Hoffer, M. I. et al. (2002) 'Advanced technology paths to global climate stability: energy for a greenhouse planet', *Science* 298 (5595): 981.

5 Lockwood, Matthew (2009) 'Climate of opinion on energy policy has changed for better', available online at www.IPPR.org (accessed 31 March 2012).

6 Hoffer et al. (2002), 981.

7 See Giddens, Anthony (2009) *The Politics of Climate Change*. Cambridge: Polity.

8 IPCC (Intergovernmental Panel on Climate Change) (2007) AR4 (Fourth Assessment Report), 'Climate Change 2007 Synthesis Report', 2, 5.

9 IFPRI (International Food Policy Research Institute) (2009) 'Climate change: impact on agriculture and costs of adaptation', available at www.ifpri.

org/publication/climate-change-impact-agriculture-and-costs-adaptation (accessed 14 March 2012).

10 Ibid., 2.

11 Funtowicz, Silvio and Jerome Ravetz (1993) 'Science from a post-normal age', *Futures* 25 (7): 739.

12 Ibid., 740.

13 Ibid., 753.

14 Charlesworth, Mark and Chukwumerije Okereke (2010) 'Policy responses to rapid climate change: an epistemological critique of dominant approaches', *Global Environmental Change* 20: 121–9.

15 Vanderheiden, Steve (2008) 'Introduction' in *Political Theory and Global Climate Change*. Cambridge MA and London: MIT Press, xiv.

16 Wynne, Brian (2010) 'Strange weather, again: climate science as political art', *Theory, Culture and Society* 27 (2–3): 289–305 and Goeminne, Gert (2012) 'Lost in translation: climate denial and the return of the political', *Global Environmental Politics* 12 (2): 1–8.

17 Hulme, Mike (2009) *Why We Disagree about Climate Change*. Cambridge: Cambridge University Press,78.

18 Hulme (2009), 82.

19 IPCC (2007), 18.

20 Committee on Climate Change (2008) 'Building a low carbon economy', available at www.theccc.org.uk/reports/building-a-low-carbon-economy (accessed 13 December 2012).

21 Risbey, James S. (2006) 'Some dangers of "dangerous" climate change', *Climate Policy* 6 (5): 527–36.

22 Jasanoff, Sheila (2007) 'Technologies of humility', *Nature* 450: 33.

23 Pielke, Roger A. (2004) 'When scientists politicise science: making sense of controversy over *The Skeptical Environmentalist*', *Environmental Science and Policy* 7: 406.

24 Hillerbrand, Rafaela and Michael Ghil (2008) 'Anthropogenic climate change: scientific uncertainties and moral dilemmas', *Physica D*. 237: 2132–8, 2137.

25 Demeritt, David (2001) 'The construction of global warming and the politics of science', *Annals of the Association of American Geographers* 91 (2): 308.

26 Bush, George W. (2007) Speech to major economies meeting on energy security and climate change, September, available at http://georgewbush-whitehouse.archives.gov/news/releases/2007/09/20070928-2.html (accessed 12 March 2012).

27 Obama, Barack (2010) 'Remarks by the President on energy' in Lanham, Maryland, February, available at www.whitehouse.gov/the-press-office/remarks-president-energy-lanham-maryland (accessed 26 July 2012).

28 Hardin, Garrett (1968) 'The tragedy of the commons', *Science* 162 (3859): 1243.

29 Ibid.

30 Ibid.

31 DECC (2011) 'The Carbon Plan: delivering our low carbon future', p. 12, available at www.decc.gov.uk/assets/decc/11/tackling-climate-change/carbon-plan/3702-the-carbon-plan-delivering-our-low-carbon-future.pdf (accessed 14 March 2012).

32 DEFRA (n.d.) 'Future world images', available at http://archive.defra.gov.uk/environment/climate/documents/interim2/future-worlds.pdf (accessed 3 March 2012).

33 Hoffer et al. (2002), 986.

34 Royal Society (2009) 'Geoengineering the climate: science, governance and uncertainty', available at http://royalsociety.org/policy/publications/2009/geoengineering-climate/ (accessed 11 March 2012), xi.

35 Ibid., v.

36 Barrett, Scott (2008) 'Climate change negotiations reconsidered', Policy Network, available online at www.policy-network.net (accessed 15 March 2011), 11.

37 Schlosberg, David and Sara Rinfret (2008) 'Ecological modernisation, American style', *Environmental Politics* 17 (2): 254–75.

38 Dryzek, John (2005b) *The Politics of the Earth: Environmental Discourses.* Oxford: Oxford University Press.

39 DECC (2011) 'The Carbon Plan: delivering our low carbon future'.

40 www.decc.gov.uk/en/content/cms/meeting_energy/renewable_ener/renewable_ener.aspx (accessed 14 March 2012).

41 Brown, Gordon (2009) 'Copenhagen or Bust' in *Newsweek*, available at www.thedailybeast.com/newsweek/2009/09/20/copenhagen-or-bust.html (accessed 26 July 2012).

42 Cameron, David (2012) Speech on Green Economy, 26 April 2012, available at www.number10.gov.uk/news/david-cameron-clean-energy/ (accessed 25 July 2012).

43 *Stern Review: The Economics of Climate Change* (2006), executive summary, vi, available at http://webarchive.nationalarchives.gov.uk/+/http://www.hm-treasury.gov.uk/stern_review_report.htm (accessed 30 March 2012).

44 Ibid., vii.

45 See, for example, Nordhaus, W. D. (2007a) 'A review of the Stern Review on the economics of climate', *Journal of Economic Literature* 45 (3): 686–702.

46 *Stern Review: The Economics of Climate Change* (2006) executive summary, ix.

47 Nordhaus (2007a), 688.

48 Ibid., 701.

49 Stern, Nicholas (2009) *Blueprint for a Safer Planet.* London: Vintage, 6.

50 Ibid., 4.

51 WBCSD (2010) 'Vision 2050: the new agenda for business' (summary), 20. Available at www.wbcsd.org/web//projects/BZrole/Vision2050_Summary_Final.pdf (accessed 27 April 2012).

52 Stern (2009), 99.

53 Ibid., 100.

54 Branson, Richard (2009) 'Response' in Tim Flannery, *Now or Never*. New York NY: Atlantic Monthly Press.

55 Schuppert, Fabian (2011) 'Climate change mitigation and intergenerational justice', *Environmental Politics* 20 (3): 303–21.

56 Nordhaus (2007a), 689.

57 Nordhaus, W. D. (2007b) 'To tax or not to tax: alternative approaches to slowing global warming', *Review of Environmental Economics and Policy* 1 (1): 26–44.

58 Pearce, David (1991) 'The role of carbon taxes in adjusting to global warming', *The Economic Journal* 101 (47): 938–48.

59 Schuppert (2011), 311.

60 Green Fiscal Commission (2010) 'Achieving fairness in carbon emissions reduction: the distributional effects of green fiscal reform', available at www.greenfiscalcommission.org.uk/ (accessed 25 July 2012).

61 Wara, M. (2007) 'Is the global carbon market working?', *Nature* 445: 595–6.

62 Convery, F. J. (2009) 'Origins and development of the EU ETS', *Environmental and Resource Economics* 43 (3): 407.

63 Tietenberg, T. 'The tradable permits approach to protecting the commons', *Oxford Review of Economic Policy* 19 (3): 400–19.

64 Schuppert (2011), 309.

65 See for example, Nordhaus (2007b), Pearce (1991), and Metcalf, G. E. (2008) 'Designing a carbon tax to reduce US greenhouse gas emissions', *Review of Environmental Economics and Policy* 3 (1): 63–83.

66 Wara (2007).

67 Reyes, Oscar and Tamra Gilbertson (2010) 'Carbon trading: how it works and why it fails', *Soundings* 45: 90.

68 Ibid., 97.

69 Ibid., 98.

70 Prudham, Scott (2009) 'Pimping the climate: Richard Branson, global warming and the performance of green capitalism', *Environment and Planning A* (41): 1595.

71 Nordhaus (2007b), 29.

72 Fawcett, Tina and Yael Parag (2010) 'An introduction to personal carbon trading', *Climate Policy* 10: 334.

73 For example, 'the world business council for sustainable development calls for governments to help it achieve its "Vision 2050"', www.wbcsd.org.

74 Jamieson, Dale (1992) 'Ethics, public policy and global warming', *Science, Technology and Human Values* 17 (2): 144.

75 Ibid., 142.

76 Ibid., 147.

77 Dobson, Andrew (2008) 'Climate change and the public sphere' in Open Democracy, available at www.opendemocracy.net (accessed 31 March 2012).

78 Titmuss, Richard (1970) *The Gift Relationship*. London: George Allen and Unwin, 245.
79 Ibid., 12.
80 Sandel, Michael (2009) *The Reith Lectures*, available at www.bbc.co.uk/programmes/b00kt7rg (accessed 31 March 2012).
81 Ibid.
82 Pearce (1991), 942.
83 Coates, John (2012) 'Unreasonable risk', *Prospect* (July).
84 Sandel, Michael (1984) 'The procedural republic and the unencumbered self', *Political Theory* 12 (1): 90.
85 Wynne (2010) 299.

Chapter 2

1 Dower, Nigel (2005) 'The nature and scope of global ethics and the relevance of the Earth Charter', *Journal of Global Ethics* 1 (1): 26.
2 See Jamieson, Dale (1992) 'Ethics, public policy and global warming', *Science, Technology and Human Values* 17 (2): 151; Gardiner, Stephen M. (2004) 'Ethics and global climate change', *Ethics* 114 (3): 555–600; Gardiner, Stephen (2006) 'A perfect moral storm: climate change, intergenerational ethics and the problem of moral corruption', *Environmental Values* 15: 397–413; Singer, Peter (2002) *One World: the Ethics of Globalisation*. New Haven CT and London: Yale University Press.
3 Gardiner (2006), 398.
4 Gardiner (2004), 595.
5 Jamieson, Dale (2008) 'The philosophers' symposium', *Critical Inquiry* 34: 612–19.
6 Jamieson (1992), 151.
7 Ibid., 151.
8 Ibid., 150.
9 Jamieson, Dale (2010) 'Climate change, responsibility and justice', *Science and Engineering Ethics* 16: 431–45.
10 Singer (2002), 19.
11 Dower, N. (2007) *World Ethics – The New Agenda*. Edinburgh: Edinburgh University Press, 7. See also Dower (2005).
12 Maniates, Michael F. (2012) 'Everyday possibilities', *Global Environmental Politics* 12 (1): 122.
13 Appiah, Kwame Anthony (2007) *Cosmopolitanism: Ethics in a World of Strangers*. London: Penguin, xiv.
14 Elliott, Lorraine (2007) 'Cosmopolitan environmental harm conventions', *Global Society* 20 (3): 350.
15 Ibid., 350.
16 Ibid., 363.

17 Harris, G. Paul (2010) *World Ethics and Climate Change*. Edinburgh: Edinburgh University Press, 118.
18 Ibid., 6.
19 Ibid., 7, emphasis in original.
20 Harris, G. Paul and Jonathan Symons (2010) 'Justice in adaptation to climate change: cosmopolitan implications for international institutions', *Environmental Politics* 19 (4): 629.
21 Ibid., 633.
22 Harris (2010), 188.
23 Caney, Simon (2005) 'Cosmopolitan justice, responsibility and global climate change', *Leiden Journal of International Law* 18: 758.
24 Ibid., 769.
25 This is because most discussions of political obligation hinge upon the absence of individual consent. There are, however, ways to justify political obligation – for example, John Horton explains that it is a sense of identification in a political community that underpins political obligation. See Horton, John (1992) *Political Obligation*. Basingstoke and London: MacMillan.
26 Dobson, Andrew (2006a) 'Thick cosmopolitanism', *Political Studies* 54 (2): 171.
27 Ibid., 171.
28 Ibid., 171.
29 Ibid., 171.
30 Ibid., 172.
31 Dobson, Andrew (2003) *Citizenship and the Environment*. Oxford and New York NY: Oxford University Press, 119
32 Dobson, Andrew (2006b) "Ecological citizenship: a defence', *Environmental Politics* 15 (3): 447.
33 Dobson (2003), 106.
34 Hayward, Tim (2006) 'Ecological citizenship: a rejoinder', *Environmental Politics* 15 (3): 452.
35 See www.defra.gov.uk/environment/quality/local/street-litter/carrier-bags/ (accessed 22 April 2012).
36 Pearce, David (1991) 'The role of carbon taxes in adjusting to global warming', *The Economic Journal* 407 (101): 939.
37 www.decc.gov.uk/en/content/cms/meeting_energy/microgen/microgen.aspx (accessed 22 April 2012).
38 Ibid.
39 Gore, Al (2006) *An Inconvenient Truth: the Planetary Emergency of Global Warming and What We Can Do About It*. London: Bloomsbury.
40 Cuomo, Chris (2011) 'Climate change, vulnerability and responsibility', *Hypatia* 26 (4): 700.
41 Note that other sources suggest that for the UK household consumption and personal transportation produce closer to 44 per cent of GHG emissions. However, regardless of the proportion, the point remains that individuals

have little control over the sources of this energy.

42 Cuomo (2011), 702.

43 Maniates (2012), 122.

44 Maniates, Michael F. (2001) 'Individualization: plant a tree, buy a bike, save the world?', *Global Environmental Politics* 1 (3): 32, emphasis in original.

45 Ibid., 38.

46 Ibid., 34.

47 G2 supplement (2010) *The Guardian*, 1 January, 1.

48 www.1010global.org/global/about (accessed 22 April 2012).

49 Katz, Ian (2009) '10:10 – The time for action', *The Guardian*, available at www.guardian.co.uk/environment/2009/dec/31/10-10-copenhagen (accessed 26 July 2012).

50 Dobson, Andrew (2009) '10:10 and the politics of climate change', *Open Democracy*, available at www.opendemocracy.net/article/10-10-and-the-politics-of-climate-change (accessed 22 April 2012).

51 Gardiner (2006), 398.

52 Ibid., 408.

Chapter 3

1 Shepard, Anna (2009) 'A case of green fatigue', *Prospect* (November), available at www.prospectmagazine.co.uk/magazine/a-case-of-green-fatigue/ (accessed 17 July 2012), 16.

2 Dobson, Andrew and Derek Bell (2006) 'Introduction' in A. Dobson and D. Bell (eds), *Environmental Citizenship*. Cambridge MA: MIT Press, 4.

3 The Green Party Manifesto, available at www.greenparty.org.uk/policies/policies_2010/2010manifesto_government.html (accessed 24 May 2012).

4 Phillips, Anne (1991) 'Citizenship and feminist politics' in Andrews, Geoff (ed.), *Citizenship*. London: Lawrence and Wishart, 81.

5 Dobson, Andrew and Angel Valencia Saiz (2005) 'Introduction', *Environmental Politics* 14 (2): 163–78.

6 For a good overview of this work on green citizenship see Dean, Hartley (2001) 'Green citizenship', *Social Policy and Administration* 35 (5): 490–505. For my critique of green citizenship see Machin, Amanda (2012) 'Decisions, disagreement and responsibility: towards an agonistic green citizenship', *Environmental Politics* 21 (6) 847–63.

7 Dobson and Bell (2006), 5.

8 Marshall, T. H. (1973) 'Citizenship and social class' in *Class, Citizenship and Social Development*. Westport CT: Greenwood Press.

9 Miller, David (2000) *Citizenship and National Identity*. Cambridge: Polity, 45.

10 Kymlicka, Will and Wayne Norman (1994) 'Return of the citizen: a survey of recent work on citizenship theory', *Ethics* 104 (2): 354.

11 Ibid., 368.

12 Smith, Mark, J. and Piya Pangsapa (2008) *Environment and Citizenship: Integrating Justice, Responsibility and Civic Engagement*. London and New York NY: Zed Books, 9.

13 Miller (2000), 53.

14 Aristotle (1912) *The Politics: A Treatise of Government*. London and Toronto: JM Dent and Sons, Book I, Chapter II, 1253a.

15 Ibid., 1261a Book II, Chapter I.

16 Held, David (1996) *Models of Democracy*. Cambridge: Polity, 36.

17 Kymlicka and Norman (1994), 362.

18 White, Stuart (2010) 'The future left: red, green and republican' in *Red Pepper*, available at www.redpepper.org.uk/the-future-left-red-green-and/ (accessed 16 May 2012).

19 Ibid.

20 Barry, John (2012) *The Politics of Actually Existing Unsustainability: Human Flourishing in a Climate-Change, Carbon-Constrained World*. Oxford: Oxford University Press, 2.

21 Ibid., 34.

22 Ibid., 260.

23 Barry, John (2008) 'Towards a green republicanism: constitutionalism, political economy, and the green state', *The Good Society* 17 (1): 6.

24 Barry (2012), 222.

25 Curry, Patrick (2000) 'Redefining community: towards an ecological republicanism', *Biodiversity and Conservation* 9: 1067.

26 Ibid., 1070.

27 Mouffe, Chantal (1992) 'Democratic citizenship and the political community' in Chantal Mouffe (ed.), *Dimensions of Radical Democracy*. London and New York NY: Verso, 227.

28 Ibid., 229.

29 Lorenzoni, Irene and Nick F. Pidgeon (2006) 'Public views on climate change: European and USA perspectives', *Climatic Change* 77: 73–95.

30 Pettit, Philip (2006) 'The republican ideal of freedom' in Miller, David (ed.), *The Liberty Reader*. Boulder CO and London: Paradigm, 229.

31 Ibid., 231.

32 Skinner, Quentin (1992) 'On justice, the common good and liberty' in Mouffe, Chantal (ed.), *Dimensions of Radical Democracy*. London and New York NY: Verso, 219.

33 Ibid., 221.

34 Miller (2000), 57.

35 Barry (2012), 7.

36 Barry, John (1999) *Rethinking Green Politics: Nature, Virtue and Progress*. London, Thousand Oaks CA and New Delhi: Sage, 203.

37 Ibid., 204.

38 Barry (2012), 12.

39 See discussion in Chapter 5.

40 Barry (1999), 232.
41 Kymlicka and Norman (1994), 360.
42 Curry (2000), 1062.
43 Barry, John (2006) 'Resistance is fertile: from environmental to sustainability citizenship' in Dobson, Andrew and Derek Bell (eds), *Environmental Citizenship*. Cambridge MA: MIT Press, 27.
44 Ibid.
45 Ibid., 33.
46 Connelly, James (2006) 'The virtues of environmental citizenship' in Dobson, A. and D. Bell, (eds), *Environmental Citizenship*. Cambridge MA: MIT Press, 65.
47 Ibid., 67.
48 Ibid., 52.
49 Hill, Thomas E. (1983) 'Ideals of human excellence and preserving natural environments', *Environmental Ethics* 5 (2): 211–24.
50 Barry (1999) and Connelly (2006).
51 Connelly (2006) 51.
52 Ibid., 51.
53 Ibid., 65.
54 Ibid., 66.
55 Lindsay, A. (1912) 'Introduction' in Aristotle, *Politics: A Treatise of Government*. London and Toronto: J. M. Dent and Sons.
56 Phillips (1991), 77.
57 Ibid., 7. See also Young, Iris (1989) 'Polity and group difference: a critique of the idea of universal citizenship', *Ethics* 99 (2): 250–74.
58 Phillips (1991), 77.
59 MacGregor, Sherilyn (2006) 'No sustainability without justice' in A. Dobson and D. Bell (eds), *Environmental Citizenship*. Cambridge MA: MIT Press, 109.
60 Lister, Ruth (1990) 'Women, economic dependency and citizenship', *Journal of Social Policy* 19 (4): 456.
61 MacGregor (2006), 109.
62 Ibid., 119.
63 Masika, Rachel (2002) 'Editorial' in *Gender and Development*. 10 (2) 2-9.
64 MacGregor, Sherilyn (2009) 'A stranger silence still: the need for feminist social research on climate change', *Sociological Review* 57: 136.
65 Terry, Geraldine (2009) 'No climate justice without gender justice' in *Gender and Development*. 17 (1): 5–18.
66 Barry (2012), 270.
67 Hayward, Bronwyn (2012) *Children, Citizenship and the Environment*. London and New York NY: Routledge, 3.
68 Wilby, Peter (2011) 'Without talk of the common good, Earth will have to fry', *The Guardian*, 3 December, available at www.guardian.co.uk/commentisfree/2011/dec/02/george-osborne-planet-durban-failure (accessed

28 May 2012).

69 See Young (1989) and Yuval-Davis, Nira (1997) *Gender and Nation*. London: Sage.

Chapter 4

1 Plato (1961) *The Republic Book V* in Hamilton, Edith and Cairns, Huntington (eds), *The Collected Dialogues of Plato*. Princeton NJ: Princeton University Press, 473 C.

2 Beck, Ulrich (2010) 'Climate for change, or how to create a green modernity', *Theory, Culture and Society* 27 (2–3): 254–66.

3 Toynbee, Polly (2009) 'Gutless, yes. But the planet's future is no priority of ours', *The Guardian*, 19 December, 35.

4 See Pendleton, Andrew (2010) 'After Cancun: shifting climate gears', available online at www.IPPR.org (accessed 15 March 2011); and Calder, Gideon and Catriona McKinnon (2011) 'Climate change and liberal priorities', *Critical Review of International Social and Political Philosophy* 14 (2): 91–7.

5 Giddens, Anthony (2008) 'The politics of climate change: national responses to the challenge of global warming', Policy Network paper, London, 8.

6 Ibid., 10.

7 Ibid., 10.

8 Ibid., 9.

9 Giddens, Anthony (2009) *The Politics of Climate Change*. Cambridge: Polity, 93.

10 Compston, Hugh and Ian Bailey (2008) 'The politics of climate policy in affluent democracies', Policy Network paper, available online at www.policy-network.net (accessed 15 March 2011), 9.

11 Hine, Dougald (2007) 'Climate change: a question of democracy', Open Democracy, available online at www.opendemocracy.net (accessed 4 June 2012).

12 Hardin, Garrett (1968) 'The tragedy of the commons', *Science* 162 (3859): 1243–8.

13 See Dobson, Andrew (2010) 'James Lovelock: greenery vs democracy', Open Democracy, available online at www.opendemocracy.net (accessed 4 June 2012).

14 James Lovelock, quoted in *The Guardian*, 29 March 2010.

15 Lovelock, James (2009) *The Vanishing Face of Gaia: a Final Warning*. London: Allen Lane, 946 and 1154.

16 Shearman, David and Joseph Wayne Smith (2007) *The Climate Challenge and the Failure of Democracy*. Westport CT: Praeger, xiii.

17 Ibid., xv.

18 Beeson, Mark (2010) 'The coming of environmental authoritarianism',

Environmental Politics 19 (2): 289.

19 See Payne, Rodger (1995) 'Freedom and the environment', *Journal of Democracy* 6 (3): 41–55; and the empirical studies of Li, Quan (2006) 'Democracy and environmental degradation', *International Studies Quarterly* 50: 935–56 and Neumayer, Eric (2002) 'Do democracies exhibit stronger international environmental commitment? A cross-country analysis', *Journal of Peace Research* 39 (2): 139–64.

20 Held, David and Angus Fane Hervey (2009) 'Democracy, climate change and global governance', Policy Network paper, available online at www.policy-network.net (accessed 15 March 2011).

21 Li (2006).

22 Ibid.

23 Payne (1995), 44.

24 Neumayer (2002), 141.

25 Democracy can be understood as an 'essentially contested' concept: 'there are concepts which are essentially contested, concepts the proper use of which inevitably involves endless disputes about their proper uses on the part of their users'. Gallie, W. B. (1956) 'Essentially contested concepts', *Proceedings of the Aristotelian Society* 56: 169.

26 Plumwood, Val (1995) "Has democracy failed ecology? An ecofeminist perspective', *Environmental Politics* 4 (4): 141.

27 Ibid., 147.

28 Taylor, Bob Pepperman (1996) 'Democracy and environmental ethics' in Lafferty, William M. and Meadowcroft, James (eds), *Democracy and the Environment*. Cheltenham and Lyme CT: Edward Elgar, 101.

29 Plumwood, Val (1993) *Feminism and the Mastery of Nature*. London: Routledge.

30 Held and Hervey (2009), 5.

31 Ibid., 6.

32 Taylor (1996), 101.

33 Dryzek, John, S. (2000) *Deliberative Democracy and Beyond: Liberals, Critics, Contestations*. Oxford: Oxford University Press,1.

34 Smith, Graham (2003) *Deliberative Democracy and the Environment*. London and New York NY: Routledge, 53.

35 Habermas, Jurgen (1994) 'Struggles for recognition in the democratic constitutional state', Shierry Weber Nicholsen (trans.) in Gutmann, Amy (ed.), *Multiculturalism*. Princeton NJ: Princeton University Press, 135.

36 Habermas, Jurgen (1996) 'Three normative models of democracy' in Behabib, Seyla (ed.), *Democracy and Difference*. Princeton NJ: Princeton University Press.

37 Habermas, Jurgen (1992) 'Citizenship and national identity: some reflections on the future of Europe', *Praxis International* 12 (1): 11.

38 Ibid., 17.

39 Benhabib, Seyla (1996) 'Toward a deliberative model of democratic legitimacy'

in S. Behabib (ed.), *Democracy and Difference*. Princeton NJ: Princeton University Press, 71.

40 Habermas, Jurgen (2002) *The Inclusion of the Other*. Cambridge: Polity Press, 41.

41 Benhabib (1996), 71.

42 Held and Hervey (2009), 8.

43 Ibid., 8.

44 Habermas (2002), 44.

45 Smith (2003), 59.

46 Ibid., 63.

47 Benhabib (1996), 73.

48 Barry, John (1999) *Rethinking Green Politics: Nature, Virtue and Progress*. London, Thousand Oaks CA and New Delhi: Sage, 229.

49 Ibid., 230.

50 Ibid., 63.

51 Dryzek (2000), 153.

52 Ibid., 140.

53 Ibid., 149.

54 Benhabib (1996), 69.

55 Young, Iris (1996) 'Communication and the other: beyond deliberative democracy' in Behabib, Seyla (ed.), *Democracy and Difference*. Princeton: Princeton University Press, 126.

56 Ibid., 123.

57 Dryzek (2000), 72.

58 Ibid., 168.

59 Coole, Diana (1996) 'Habermas and the question of alterity' in d'Entrèves, Maurizio and Syla Benhabib (eds), *Habermas and the Unfinished Project of Modernity*. Cambridge: Polity Press, 226.

60 Ibid., 226.

61 Ibid., 227.

62 Ibid., 225.

63 Vidal, John (2009) 'Lifting the lid on climate talks', *The Guardian*, available at www.guardian.co.uk/environment/2009/nov/07/climate-change-talks-2009 (accessed 26 July 2012).

64 Dryzek (2000), 146.

65 Ibid., 148.

66 Ibid., 2.

67 Tully, James (1989) 'Wittgenstein and political philosophy', *Political Theory* 17 (2): 181.

68 Tully, James (2001) 'An ecological ethics for the present' in Gleeson, Brendan and Nicholas Low (eds), *Governing for the Environment: Global Problems, Ethics and Democracy*. Basingstoke: Palgrave, 149.

69 Ibid., 151.

70 Habermas (2002), 40.

71 Mouffe, Chantal (1999) 'Deliberative democracy or agonistic pluralism', *Social Research* 66 (3): 756.
72 Mouffe, Chantal (2000) *The Democratic Paradox*. London and New York NY: Verso, 98.

Chapter 5

1 Hulme, Mike (2009) *Why We Disagree about Climate Change*. Cambridge: Cambridge University Press, 357.
2 Plumwood, Val (1995) 'Has democracy failed ecology? An ecofeminist perspective', *Environmental Politics* 3 (2): 134–68.
3 Giddens, Anthony (2009) *The Politics of Climate Change*. Cambridge: Polity, 2.
4 Collins English Dictionary.
5 Zizek, Slavoj (2009) *The Parallax View*. Cambridge MA: MIT Press, 4.
6 Mouffe, Chantal (2000) *The Democratic Paradox*. London and New York NY: Verso, 15.
7 Ibid., 137.
8 Rancière, Jacques (1999) *Disagreement*. Minneapolis MN and London: University of Minnesota Press, 102. See also Rancière, Jacques (2001) 'The theses on politics', *Theory and Event* 5 (3).
9 Rancière (1999), 101.
10 Ibid., 116.
11 Ibid., 30.
12 Dikec, Mustafa (2005) 'Space, politics and the political', *Environment and Planning D: Society and Space* 23: 176.
13 Ibid., 177.
14 See Tambakaki, Paulina (2009) 'When does politics happen?' in *Parallax* 52.
15 Mouffe (2000), 13.
16 Mouffe, Chantal (2005) *On The Political*. London and New York NY: Routledge, 21.
17 Mouffe (2005), 4.
18 Mouffe (2000), 113.
19 Knight, Sam (2009) 'Eleven days in December' in *Prospect*, 21 October. Available at www.prospectmagazine.co.uk/magazine/eleven-days-in-december/ (accessed 21 July 2012), 3.
20 Ibid., 8.
21 Giddens (2009), 7.
22 IPPR (2006) Ereaut, Gill and Segnit, Nat, 'Warm words: how are we telling the climate story and can we tell it better?' Available at www.ippr.org/images/media/files/publication/2011/05/warm_words_1529.pdf (accessed 23 July 2012). See also Gillard, Julia (2010) 'Moving forward together on climate change', speech at the University of Queensland. Available at www.alp.org.au/federal-government/news/speech--julia-gillard,--moving-forward-

together-on/ (accessed 3 July 2012) and Stern, Nicholas (2009) 'The rich the poor and the planet' in LSE Connect, available at www2.lse.ac.uk/ GranthamInstitute/Media/Articles/rich-poor-planet-NY-stern.pdf (accessed 23 July 2012).

23 Wynne, Brian (2010) 'Strange weather, again: climate science as political art', *Theory, Culture and Society* 27 (2–3): 295.

24 Ibid., 300.

25 Goeminne, Gert (2012) 'Lost in translation: climate denial and the return of the political', *Global Environmental Politics* 12 (2): 6.

26 Ibid., 6.

27 Swyngedouw, Erik (2010) 'Apocaypse forever? Post-political populism and the spectre of climate change', *Theory, Culture and Society* 27 (2–3): 228.

28 Aitken, Mhairi (2012) 'Changing climate, changing democracy: a cautionary tale', *Environmental Politics* 21 (2): 219.

29 Ibid., 225.

30 See also Hulme (2009) and Hayward, Bronwyn (2008) 'Let's talk about the weather: decentering democratic debate about climate change', *Hypatia* 23 (3): 79–98.

31 Hutton, Will (2012) 'Global warming off the agenda? Now that would be a catastrophe', *The Observer*, 24 June.

32 Rancière, Jacques (2004) 'Introducing disagreement', *Angelaki* 9 (3): 7.

33 Mouffe (2000), 7.

34 Goeminne (2012), 7.

35 Klein, Naomi (2011) 'Capitalism vs climate', *The Nation*, 28 November 2011, available at www.thenation.com/article/164497/capitalism-vs-climate (accessed 12 July 2012), 1.

36 Mouffe (2005), 5.

37 Juniper, Tony (2006) 'Beyond all reasonable doubt', *The Guardian*, 3 November, available at www.guardian.co.uk/commentisfree/2006/nov/03/ post572 (accessed 23 July 2012). See also *Nature* (2010) 'Turbines and turbulence', *Nature* 258: 1001, cited by Aitken (2012). Journal name:

38 Klein (2011), 4.

39 Smith, Graham (2003) *Deliberative Democracy and the Environment*. London and New York NY: Routledge, 3.

40 Ibid., 5.

41 Ibid., 60.

42 Barry, John and Geraint Ellis (2010) 'Beyond consensus? Agonism, republicanism and a low carbon future' in Devine-Wright, P. (ed.), *Renewable Energy and the Public*. London: Earthscan, 10.

43 Ibid., 6.

44 Ibid., 15.

45 Ibid., 3.

46 Ibid., 9.

47 Rancière (1999), x.

48 Dryzek, John (2005a) 'Deliberative democracy in divided societies: alternatives to agonism and analgesia', *Political Theory* 33 (2): 221.

49 Ibid., 226.

50 Mouffe (2000), 131.

51 Mouffe (2000), 136.

52 Latta, Alex (2007), 'Locating democratic politics in ecological citizenship', *Environmental Politics* 16 (3): 379.

53 Ibid., 379.

54 Ibid., 385.

55 Torgerson (1999), 19.

56 Torgerson (2000), 7.

57 Ibid., 16.

58 Smith (2003), 72.

59 Edward Davey, speech to the Global Offshore Wind Conference, 14 June 2012, available at www.decc.gov.uk/en/content/cms/news/edd_globaloffw/edd_globaloffw.aspx (accessed 21 July 2012).

60 Environmental Audit Committee (2007), 'Beyond Stern: from the Climate Change Programme Review to the Draft Climate Change Bill'. London: The Stationery Office.

61 Rootes, Christopher and Neil Carter (2010) 'Take blue, add yellow, get green? The environment in the UK general election of 6 May 2010', *Environmental Politics* 19 (6): 992–9.

62 Mouffe (2000), 105.

Chapter 6

1 Lovelock, James (2008) 'Forward' in Henson, Robert, *The Rough Guide to Climate Change*. London: Rough Guides, vii.

2 Flannery, Tim (2009) *Now or Never: Why We Must Act Now to End Climate Change and Create a Sustainable Future*. New York NY: Atlantic Monthly Press, 14.

3 Gore, Al (2006) *An Inconvenient Truth: the Planetary Emergency of Global Warming and What We Can Do about It*. London: Bloomsbury, back cover.

4 For example, *An Inconvenient Truth* (2006), film directed by Davis Guggenheim, Lawrence Bender Productions (USA); and *The Age of Stupid* (2009), film directed by Franny Armstrong, Spanner Films (UK).

5 Novels such as Atwood, Margaret (2009) *The Year of the Flood*. London, Berlin and New York NY: Bloomsbury; McEwan, Ian (2011) *Solar*. London: Vintage; McCarthy, Cormac (2006) *The Road*. London: Picador. Films such as *The Day after Tomorrow* (2004), directed by Roland Emmerich, Twentieth Century Fox; and theatre productions such as *Greenland* (2011), play directed by Bijan Sheibani, National Theatre, London.

6 Knight, Sam (2009) 'Eleven days in December', *Prospect*, 21 October,

available at www.prospectmagazine.co.uk/magazine/eleven-days-in-december/ (accessed 21 July 2012), 3.

7 See Wynne, Brian (2010) 'Strange weather, again: climate science as political art', *Theory, Culture and Society* 27 (2–3): 289–305; Hayward, Bronwyn (2008) 'Let's talk about the weather: decentering democratic debate about climate change', *Hypatia* 23 (3): 79–98.

8 Torgerson, Douglas (1999) *The Promise of Green Politics: Environmentalism and the Public Sphere*. Durham NC and London: Duke University Press; Dryzek, John, S. (2000) *Deliberative Democracy and Beyond: Liberals, Critics, Contestations*. Oxford: Oxford University Press.

9 Grosz, Elizabeth (1994) *Volatile Bodies: Towards a Corporeal Feminism*. Bloomington and Indianapolis IN: Indiana University Press, 3.

10 Plumwood, Val (1993) *Feminism and the Mastery of Nature*. London and New York NY: Routledge, 3.

11 Katz, Cindy and Andrew Kirby (1991) 'In the nature of things: the environment and everyday life', *Transactions of the Institute of British Geographers* 16 (3): 265.

12 Katz and Kirby (1991); see also Torgerson, Douglas (2006) 'Expanding the green public sphere: post-colonial connections', *Environmental Politics* 15 (5): 713–30.

13 Soper, Kate (2000) 'Future culture: realism, humanism and the politics of nature', *Radical Philosophy* 120: 18.

14 Soper, Kate (1995) *What Is Nature?* Oxford and Malden: Wiley-Blackwell, 2.

15 Soper (2000), 18.

16 Ibid., 22.

17 Young, Iris (2005) *On Female Body Experience: 'Throwing Like a Girl' and Other Essays*. Oxford: Oxford University Press, 29.

18 Gabrielson, Teena and Katelyn Parady (2010) 'Green citizenship: rethinking green citizenship through the body', *Environmental Politics* 19 (3): 386.

19 Hulme, Mike (2009) *Why We Disagree about Climate Change*. Cambridge: Cambridge University Press, 4.

20 Ibid., 33.

21 Torgerson (2006) and Smith, Graham (2003) *Deliberative Democracy and the Environment*. London and New York NY: Routledge.

22 Dryzek, John (2005) *The Politics of the Earth: Environmental Discourses*. Oxford: Oxford University Press, 234 .

23 Kingsnorth, Paul (2010) 'Confessions of a recovering environmentalist' on *OpenDemocracy*, available at www.opendemocracy.net/paul-kingsnorth/confessions-of-recovering-environmentalist (accessed 25 June 2012).

24 Ibid.

25 Katz and Kirby (1991).

26 Calhoun, Craig (2004) 'A world of emergencies: fear, intervention, and the limits of the cosmopolitan order', *Canadian Review of Sociology and Anthropology* 41 (4): 374.

27 Ibid., 375.

28 Kingsnorth, Paul (2009) 'A climate deal is like trying to halt the rains in Cumbria', *The Guardian*, 25 November.

29 Gore (2006), 3.

30 Swyngedouw, Erik (2010) 'Apocalypse forever? Post-political populism and the spectre of climate change', *Theory, Culture and Society* 27 (2–3): 216.

31 Ibid., 221.

32 Wynne (2010), 295.

33 Hammond, Philip and Hugh Ortega Breton (forthcoming) 'Bridging the political deficit: loss, morality and agency in films addressing climate change', *Communication, Culture and Critique*.

34 Ibid., 10.

35 Gardiner, Stephen, M. (2004) 'Ethics and global climate change, *Ethics* 114 (3): 595.

36 Heath, Malcolm (1996) 'Introduction' in *Aristotle: Poetics*. London: Penguin.

37 Torgerson, Douglas (2000) 'Farewell to the green movement? Political action and the green public sphere', *Environmental Politics* 9 (4): 84.

38 Ibid., 87.

39 Ibid., 103.

40 Ibid., 89.

41 Swyngedouw (2010), 219.

42 Ibid., 223.

43 Ibid., 226.

44 Wynne (2010).

45 Hulme (2009), 325.

46 Ibid., xxvi.

47 Ibid., 333.

48 Ibid., 326.

49 Ibid., 364.

50 Swyngedouw (2009), 614.

51 Monbiot, George (2009) 'This is bigger than climate change. It is a battle to redefine humanity', *The Guardian*, 14 December, available at www.guardian.co.uk/commentisfree/cif-green/2009/dec/14/climate-change-battle-redefine-humanity (accessed 26 July 2012).

52 Klein, Naomi (2011) 'Capitalism vs climate', *The Nation*, 28 November 2011, available at www.thenation.com/article/164497/capitalism-vs-climate (accessed 12 July 2012), 3.

53 Connelly, William E. (1991) *Identity/Difference: Democratic Negotiations of Political Paradox*. Minneapolis MN and London: University of Minnesota Press, xiv.

54 Ibid., 64.

55 Laclau, Ernesto and Chantal Mouffe (2001) *Hegemony and Socialist Strategy*. London and New York NY: Verso, 125.

56 Klein (2011), 4.

57 Giddens, Anthony (2009) *The Politics of Climate Change*. Cambridge, Polity, 6.

58 Rorty, Richard (1989) *Contingency, Irony and Solidarity*. Cambridge: Cambridge University Press, 191.

59 World Resources Institute (2005) 'Navigating the numbers', available at http://pdf.wri.org/navigating_numbers.pdf (accessed 25 July 2012), viii.

60 Parker, Charles F. et al. (2012) 'Fragmented climate change leadership: making sense of the ambiguous outcome of COP-15', *Environmental Politics* 21 (20): 268–86.

61 Dimitrov, Radoslav, D. (2010) 'Inside Copenhagen: the state of climate governance', *Global Environmental Politics* 10 (2): 18.

62 Hutton, Will (2012) 'Global warming off the agenda? Now that would be a catastrophe', *The Observer,* 24 June.

63 Parker et al. (2012).

64 See Pendleton (2009) 'After Copenhagen' on opendemocracy, available at www.opendemocracy.net/andrew-pendleton/after-copenhagen (accessed 26 July 2012); Christoff, Peter (2010) 'Cold climate in Copenhagen: China and the United States at COP15', *Environmental Politics* 19 (4): 637–56.

65 Barrett, Scott, (2008) 'Climate change negotiations reconsidered', Policy Network, available online at www.policy-network.net (accessed 15 March 2011).

66 Hayward (2008), 87.

67 Ibid., 87.

68 Ibid., 88.

Bibliography

Aitken, Mhairi (2012) 'Changing climate, changing democracy: a cautionary tale', *Environmental Politics* 21 (2): 211–29.

An Inconvenient Truth (2006) Film directed by Davis Guggenheim, Lawrence Bender Productions, USA.

Appiah, Kwame Anthony (2007) *Cosmopolitanism: Ethics in a World of Strangers*. London: Penguin.

Aristotle (1912) *Politics: A Treatise of Government*. London and Toronto: J. M. Dent and Sons Ltd.

Arrhenius, Svante (1896) 'On the influence of carbonic acid in the air upon the temperature of the ground', *Philosophical Magazine and Journal of Science* 41: 237–76.

Atwood, Margaret (2009) *The Year of the Flood*. London, Berlin and New York NY: Bloomsbury.

Barrett, Scott (2008) 'Climate change negotiations reconsidered', Policy Network, available online at www.policy-network.net (accessed 15 March 2011).

Barry, John (1999) *Rethinking Green Politics: Nature, Virtue and Progress*. London, Thousand Oaks CA and New Delhi: Sage.

—— (2006) 'Resistance is fertile: from environmental to sustainability citizenship' in Dobson, Andrew and Derek Bell (eds), *Environmental Citizenship*. Cambridge MA: MIT Press.

—— (2008) 'Towards a green republicanism: constitutionalism, political economy, and the green state', *The Good Society* 17 (1): 1–11.

—— (2012) *The Politics of Actually Existing Unsustainability: Human Flourishing in a Climate-Change, Carbon-Constrained World*. Oxford: Oxford University Press.

Barry, John and Ellis, Geraint (2010) 'Beyond consensus? Agonism, republicanism and a low carbon future' in Devine-Wright, P. (ed.), *Renewable Energy and the Public*. London: Earthscan.

Beck, Ulrich (2010) 'Climate for change, or how to create a green modernity', *Theory, Culture and Society* 27 (2–3): 254–66.

Beeson, Mark (2010) 'The coming of environmental authoritarianism',

Environmental Politics 19 (2): 276–94.

Benhabib, Seyla (1996) 'Toward a deliberative model of democratic legitimacy' in Behabib, Seyla (ed.), *Democracy and Difference*. Princeton: Princeton University Press.

Branson, Richard (2009) 'Response' in Flannery, Tim, *Now or Never*. New York NY: Atlantic Monthly Press.

Brown, Gordon (2009) 'Copenhagen or bust', *Newsweek*, 20 September, available at www.thedailybeast.com/newsweek/2009/09/20/copenhagen-or-bust.html (accessed 26 July 2012).

Bush, George W. (2007) 'Speech to major economies meeting on energy security and climate change', September, available at http://georgewbush-whitehouse.archives.gov/news/releases/2007/09/20070928-2.html (accessed 12 March 2012).

Calder, Gideon and Catriona McKinnon (2011) 'Climate change and liberal priorities', *Critical Review of International Social and Political Philosophy* 14 (2): 91–7.

Calhoun, Craig (2004) 'A world of emergencies: fear, intervention, and the limits of the cosmopolitan order', *Canadian Review of Sociology and Anthropology* 41 (4:) 373–95.

Cameron, David (2012) Speech on green economy, 26 April 2012, available at www.number10.gov.uk/news/david-cameron-clean-energy/ (accessed 25 July 2012).

Caney, Simon (2005) 'Cosmopolitan justice, responsibility and global climate change', *Leiden Journal of International Law* 18: 747–5.

Charlesworth, Mark and Chukwumerije Okereke (2010) 'Policy responses to rapid climate change: an epistemological critique of dominant approaches', *Global Environmental Change* 20: 121–9.

Christoff, Peter (2010) 'Cold climate in Copenhagen: China and the United States at COP15', *Environmental Politics* 19 (4): 637–56.

Coates, John (2010) 'Unreasonable risk', *Prospect* (July).

Committee on Climate Change (2008) 'Building a low carbon economy', available at www.theccc.org.uk/reports/building-a-low-carbon-economy (accessed 13 December 2012).

Compston, Hugh and Ian Bailey (2008) 'The politics of climate policy in affluent democracies', Policy Network, available online at www.policy-network.net (accessed 15 March 2011).

Connelly, James (2006) 'The virtues of environmental citizenship' in Dobson, A. and D. Bell. (eds), *Environmental Citizenship*. Cambridge MA: MIT Press.

Connelly, William E. (1991) *Identity/Difference: Democratic Negotiations of Political Paradox*. Minneapolis MN and London: University of Minnesota Press.

Convery, F. J. (2009) 'Origins and development of the EU ETS', *Environmental and Resource Economics* 43 (3): 391–412.

Coole, Diana (1996) 'Habermas and the question of alterity' in d'Entrèves, Maurizio and Syla Benhabib (eds), *Habermas and the Unfinished Project of*

Modernity. Cambridge: Polity Press.

Cuomo, Chris (2011) 'Climate change, vulnerability and responsibility', *Hypatia* 26 (4): 690–714.

Curry, Patrick (2000) 'Redefining community: towards an ecological republicanism', *Biodiversity and Conservation* 9: 1059–71.

Dean, Hartley (2001) 'Green citizenship', *Social Policy and Administration* 35 (5): 490–505.

DECC (2011) 'The Carbon Plan: delivering our low carbon future', 12. Available at www.decc.gov.uk/assets/decc/11/tackling-climate-change/carbon-plan/3702-the-carbon-plan-delivering-our-low-carbon-future.pdf (accessed 14 March 2012).

DEFRA (n.d.) 'Future world images', available at http://archive.defra.gov.uk/environment/climate/documents/interim2/future-worlds.pdf (accessed 3 March 2012).

Demeritt, David (2001) 'The construction of global warming and the politics of science', *Annals of the Association of American Geographers* 91 (2): 307–37.

Dikec, Mustafa (2005) 'Space, politics and the political', *Environment and Planning D: Society and Space* 23: 171–88.

Dimitrov, Radoslav, D. (2010) 'Inside Copenhagen: the state of climate governance', *Global Environmental Politics* 10 (2): 18–24.

Dobson, Andrew (2003) *Citizenship and the Environment*. Oxford and New York NY: Oxford University Press.

—— (2006a) 'Thick cosmopolitanism', *Political Studies* 54 (2): 165–84.

—— (2006b) 'Ecological citizenship: a defence', *Environmental Politics* 15 (3): 447–51.

—— (2008) 'Climate change and the public sphere', Open Democracy, available at www.opendemocracy.net (accessed 31 March 2012).

—— (2009) '10:10 and the politics of climate change', Open Democracy, available at www.opendemocracy.net/article/10-10-and-the-politics-of-climate-change (accessed 22 April 2012).

—— (2010) 'James Lovelock: greenery vs democracy', Open Democracy, available online at www.opendemocracy.net (accessed 4 June 2012).

Dobson, Andrew and Derek Bell (2006). 'Introduction' in Dobson, A. and Bell, D. (eds), *Environmental Citizenship*. Cambridge, MA: MIT Press.

Dobson, Andrew and Angel Valencia Saiz (2005) 'Introduction', *Environmental Politics* 14 (2): 163–78.

Dower, Nigel (2005) 'The nature and scope of global ethics and the relevance of the Earth Charter', *Journal of Global Ethics* 1 (1): 25–43.

—— (2007) *World Ethics – the New Agenda*. Edinburgh: Edinburgh University Press.

Dryzek, John, S. (2000) *Deliberative Democracy and Beyond: Liberals, Critics, Contestations*. Oxford: Oxford University Press.

—— (2005a) 'Deliberative democracy in divided societies: alternatives to agonism and analgesia', *Political Theory* 33 (2): 218–42.

—— (2005b) *The Politics of the Earth: Environmental Discourses*. Oxford: Oxford University Press.

Dunlap, Thomas R. (2004) *Faith in Nature: Environmentalism as Religious Quest*. Seattle WA and London: University of Washington Press.

Elliott, Lorraine (2007) 'Cosmopolitan environmental harm conventions', *Global Society* 20 (3): 345–63.

Environmental Audit Committee (2007) 'Beyond Stern: from the Climate Change Programme Review to the Draft Climate Change Bill'. London: The Stationery Office.

Fawcett, Tina and Yael Parag (2010) 'An introduction to personal carbon trading', *Climate Policy* 10: 329–38.

Flannery, Tim (2009) *Now or Never: Why We Must Act Now to End Climate Change and Create a Sustainable Future*. New York NY: Atlantic Monthly Press.

Funtowicz, Silvio and Jerome Ravetz (1993) 'Science from a post-normal age', *Futures* 25 (7): 739–55.

Gabrielson, Teena and Katelyn Parady (2010) 'Green citizenship: rethinking green citizenship through the body', *Environmental Politics* 19 (3): 374–91.

Gallie, W. B. (1956) 'Essentially contested concepts', *Proceedings of the Aristotelian Society* 56: 167–98.

Gardiner, Stephen, M. (2004) 'Ethics and global climate change', *Ethics* 114 (3): 555–600.

—— (2006) 'A perfect moral storm: climate change, intergenerational ethics and the problem of moral corruption', *Environmental Values* 15: 397–413.

Giddens, Anthony (2008) *The Politics of Climate Change: National Responses to the Challenge of Global Warming*. London: Policy Network.

—— (2009) *The Politics of Climate Change*. Cambridge, Polity.

Gillard, Julia (2010) 'Moving forward together on climate change', speech to University of Queensland, available at www.alp.org.au/federal-government/news/speech--julia-gillard,--moving-forward-together-on/ (accessed 3 July 2012).

Goeminne, Gert (2012) 'Lost in translation: climate denial and the return of the political', *Global Environmental Politics* 12 (2): 1–8.

Gore, Al (2006) *An Inconvenient Truth: the Planetary Emergency of Global Warming and What We Can Do about It* London: Bloomsbury.

—— (2006) *An Inconvenient Truth* DVD, Paramount Home Entertainment.

Green Fiscal Commission (2010) 'Achieving fairness in carbon emissions reduction: the distributional effects of green fiscal reform', available at www.greenfiscalcommission.org.uk/ (accessed 25 July 2012).

Greenland (2011) Play directed by Bijan Sheibani, National Theatre, London.

Grosz, Elizabeth (1994) *Volatile Bodies: Towards a Corporeal Feminism*. Bloomington and Indianapolis IN: Indiana University Press.

Habermas, Jurgen (1992) 'Citizenship and national identity: some reflections on the future of Europe', *Praxis International* 12 (1): 1–19.

—— (1994) 'Struggles for recognition in the democratic constitutional state', Shierry Weber Nicholsen (trans.), in Gutmann, Amy (ed.), *Multiculturalism*. Princeton NJ: Princeton University Press.

—— (1996) 'Three normative models of democracy' in Behabib, Seyla (ed.), *Democracy and Difference*. Princeton NJ: Princeton University Press.

—— (2002) *The Inclusion of the Other*. Cambridge: Polity Press.

Hammond, Philip and Hugh Ortega Breton (forthcoming) 'Bridging the political deficit: loss, morality and agency in films addressing climate change', *Communication, Culture and Critique*.

Hardin, Garrett (1968) 'The tragedy of the commons', *Science* 162 (3859): 1243–8.

Harris, G. Paul (2010) *World Ethics and Climate Change*. Edinburgh: Edinburgh University Press.

Harris, G. Paul and Jonathan Symons (2010) 'Justice in adaptation to climate change: cosmopolitan implications for international institutions', *Environmental Politics* 19 (4): 617–36.

Hayward, Bronwyn (2008) 'Let's talk about the weather: decentering democratic debate about climate change', *Hypatia* 23 (3): 79–98.

—— (2012) *Children, Citizenship and the Environment*. London and New York NY: Routledge.

Hayward, Tim (2006) 'Ecological citizenship: a rejoinder', *Environmental Politics* 15 (3): 452–3.

Heath, Malcolm (1996) 'Introduction' in *Aristotle: Poetics*. London: Penguin.

Held, David (1996) *Models of Democracy*. Cambridge: Polity.

Held, David and Angus Fane Hervey (2009) 'Democracy, climate change and global governance', Policy Network, available online at www.policy-network.net (accessed 15 March 2011).

Hill, Thomas E. (1983) 'Ideals of human excellence and preserving natural environments', *Environmental Ethics* 5 (2): 211–24.

Hillerbrand, Rafaela and Michael Ghil (2008) 'Anthropogenic climate change: scientific uncertainties and moral dilemmas', *Physica D*. 237: 2132–8.

Hine, Dougald (2007) 'Climate change: a question of democracy', Open Democracy, available online at www.opendemocracy.net (accessed 4 June 2012).

Hoffer, M. I. et al. (2002) 'Advanced technology paths to global climate stability: energy for a greenhouse planet', *Science* 298 (5595): 981.

Horton, John (1992) *Political Obligation*. Basingstoke and London: MacMillan.

Hulme, Mike (2009) *Why We Disagree about Climate Change*. Cambridge: Cambridge University Press.

Hutton, Will (2012) 'Global warming off the agenda? Now that would be a catastrophe', *The Observer*, 24 June.

IFPIR (2009) 'Climate change: impact on agriculture and costs of adaptation', available at www.ifpri.org/publication/climate-change-impact-agriculture-and-costs-adaptation (accessed 14 March 2012).

IPCC (Intergovernmental Panel on Climate Change) (2007) 'Climate Change 2007: Synthesis Report'.

IPPR (2006) Ereaut, Gill and Nat Segnit, 'Warm words: how are we telling the climate story and can we tell it better?' Available at www.ippr.org/images/media/files/publication/2011/05/warm_words_1529.pdf (accessed 23 July 2012).

Jamieson, Dale (1992) 'Ethics, public policy and global warming', *Science, Technology and Human Values* 17 (2): 139–53.

—— (2008) 'The philosophers' symposium', *Critical Inquiry* 34: 612–19.

—— (2010) 'Climate change, responsibility and justice', *Science and Engineering Ethics* 16: 431–45.

Jasanoff, Shelia (2007) 'Technologies of humility', *Nature* 450: 33.

Juniper, Tony (2006) 'Beyond all reasonable doubt', *The Guardian*, 3 November, available at www.guardian.co.uk/commentisfree/2006/nov/03/post572 (accessed 23 July 2012).

Katz, Ian (2009) '10:10 – the time for action', *The Guardian*, 31 December, available at www.guardian.co.uk/environment/2009/dec/31/10-10-copen hagen (accessed 26 July 2012).

Katz, Cindy and Andrew Kirby (1991) 'In the nature of things: the environment and everyday life', *Transactions of the Institute of British Geographers* 16 (3): 259–71.

Kingsnorth, Paul (2009) 'A climate deal is like trying to halt the rains in Cumbria', *The Guardian*, 25 November.

—— (2010) 'Confessions of a recovering environmentalist', Open Democracy, available at www.opendemocracy.net/paul-kingsnorth/confessions-of-recovering-environmentalist (accessed 25 June 2012).

Klein, Naomi (2011) 'Capitalism vs climate', *The Nation*, 28 November 2011, available at www.thenation.com/article/164497/capitalism-vs-climate (accessed 12 July 2012).

Knight, Sam (2009) 'Eleven days in December', *Prospect*, 21 October, available at www.prospectmagazine.co.uk/magazine/eleven-days-in-december/ (accessed 21 July 2012).

Kymlicka, Will and Wayne Norman (1994) 'Return of the citizen: a survey of recent work on citizenship theory', *Ethics* 104 (2): 352–81.

Laclau, Ernesto and Chantal Mouffe (2001) *Hegemony and Socialist Strategy*. London and New York NY: Verso.

Latta, Alex (2007) 'Locating democratic politics in ecological citizenship', *Environmental Politics* 16 (3): 377–83.

Li, Quan (2006) 'Democracy and environmental degradation', *International Studies Quarterly* 50: 935–56.

Lindsay, A. (1912) 'Introduction' in Aristotle, *Politics: A Treatise of Government*. London and Toronto: J. M. Dent and Sons Ltd.

Lister, Ruth (1990) 'Women, economic dependency and citizenship', *Journal of Social Policy* 19 (4): 445–67.

Lockwood, Matthew (2009) 'Climate of opinion on energy policy has changed for better', available online at www.ippr.org (accessed 15 March 2011).

Lorenzoni, Irene and Nick F. Pidgeon (2006) 'Public views on climate change: European and USA perspectives', *Climatic Change* (77): 73–95.

Lovelock, James (2008) 'Forward' in Henson, Robert, *The Rough Guide to Climate Change*. London: Rough Guides.

—— (2009) *The Vanishing Face of Gaia: A Final Warning*. London: Allen Lane.

MacGregor, Sherilyn (2006) 'No sustainability without justice' in Dobson, Andrew and Derek Bell (eds), *Environmental Citizenship*. Cambridge MA: MIT Press.

—— (2009) 'A stranger silence still: the need for feminist social research on climate change', *Sociological Review* 57: 124–40.

Machin, Amanda (2012) 'Decisions, disagreement and responsibility: towards an agonistic green citizenship', *Environmental Politics* 21 (6): 847–63.

Maniates, Michael F. (2001) 'Individualization: plant a tree, buy a bike, save the world?', *Global Environmental Politics* 1 (3): 31–52.

—— (2012) 'Everyday possibilities', *Global Environmental Politics* 12 (1): 121–5.

Marshall, T. H. ([1950] 1973) 'Citizenship and social class' in *Class, Citizenship and Social Development*. Westport CT: Greenwood Press.

Masika, Rachel (2002) 'Editorial', *Gender and Development* 10 (2): 2–9.

McCarthy, Cormac (2006) *The Road*. London: Picador.

McEwan, Ian (2011) *Solar*. London: Vintage.

Metcalf, G. E. (2008) 'Designing a carbon tax to reduce US greenhouse gas emissions', *Review of Environmental Economics and Policy* 3 (1): 63–83.

Miller, David (2000) *Citizenship and National Identity*. Cambridge and Malden: Polity.

Monbiot, George (2009) 'This is bigger than climate change. It is a battle to redefine humanity', *The Guardian*, 14 December, available at www.guardian.co.uk/commentisfree/cif-green/2009/dec/14/climate-change-battle-redefine-humanity (accessed 26 July 2012).

Mouffe, Chantal (1992) 'Democratic citizenship and the political community' in Mouffe, Chantal (ed.), *Dimensions of Radical Democracy*. London and New York NY: Verso.

—— (1999) 'Deliberative democracy or agonistic pluralism', *Social Research* 66 (3): 745–58.

—— (2000) *The Democratic Paradox*. London and New York NY: Verso.

—— (2005) *On The Political*. London and New York NY: Routledge.

Nature (2010) 'Turbines and turbulence', *Nature* 258: 1001.

Neumayer, Eric (2002) 'Do democracies exhibit stronger international environmental commitment? A Cross-country analysis', *Journal of Peace Research* 39 (2): 139–64.

Nordhaus, W. D. (2007a) 'A review of the Stern Review on the economics of climate', *Journal of Economic Literature* 45 (3): 686–702.

—— (2007b) 'To tax or not to tax: alternative approaches to slowing global warming', *Review of Environmental Economics and Policy* 1 (1): 26–44.

Obama, Barack (2010) 'Remarks by the President on Energy in Lanham, Maryland', February, available at www.whitehouse.gov/the-press-office/remarks-president-energy-lanham-maryland (accessed 26 July 2012).

Oreskes, Naomi and Erik M. Conway (2010) *Merchants of Doubt*. London: Bloomsbury.

Parker, Charles F. et al. (2012) 'Fragmented climate change leadership: making sense of the ambiguous outcome of COP-15', *Environmental Politics* 21 (20): 268–86.

Payne, Rodger (1995) 'Freedom and the environment', *Journal of Democracy* (6) 41–55.

Pearce, David (1991) 'The role of carbon taxes in adjusting to global warming', *The Economic Journal* 407 (101): 938–48.

Pendleton, Andrew (2009) 'After Copenhagen', Open Democracy, available at www.opendemocracy.net/andrew-pendleton/after-copenhagen (accessed 26 July 2012).

—— (2010) 'After Cancun: shifting climate gears', available online at www.IPPR.org (accessed 15 March 2011).

Pettit, Philip (2006) 'The republican ideal of freedom' in Miller, David (ed.), *The Liberty Reader*. Boulder CO and London: Paradigm.

Phillips, Anne (1991) 'Citizenship and feminist politics' in Andrews, Geoff (ed.), *Citizenship*. London: Lawrence and Wishart.

Pielke, Roger A. (2004) 'When scientists politicise science: making sense of controversy over *The Skeptical Environmentalist*', *Environmental Science and Policy* (7): 405–17.

Plato (1961) *The Collected Dialogues of Plato*, eds Edith Hamilton and Huntington Cairns. Princeton NJ: Princeton University Press.

Plumwood, Val (1993) *Feminism and the Mastery of Nature*. London: Routledge.

—— (1995) 'Has democracy failed ecology? An ecofeminist perspective', *Environmental Politics* 4 (4): 134–68.

Prudham, Scott (2009) 'Pimping the climate: Richard Branson, global warming and the performance of green capitalism', *Environment and Planning A* (41): 1594–1613.

Rancière, Jacques (1999) *Disagreement*. Minneapolis MN and London: University of Minnesota Press.

—— (2001) 'The theses on politics', *Theory and Event* 5 (3).

—— (2004) 'Introducing disagreement', *Angelaki* 9 (3): 3–9.

Reyes, Oscar and Tamara Gilbertson (2010) 'Carbon trading: how it works and why it fails', *Soundings* (45): 89–100.

Risbey, James S. (2006) 'Some dangers of "dangerous" climate change', *Climate Policy* 6 (5): 527–36.

Rootes, Christopher and Neil Carter (2010) 'Take blue, add yellow, get green? The environment in the UK general election of 6 May 2010', *Environmental Politics* 19 (6): 992–9.

Rorty, Richard (1989) *Contingency, Irony and Solidarity*. Cambridge: Cambridge University Press.

Royal Society (2009) 'Geoengineering the climate: science, governance and uncertainty', available at http://royalsociety.org/policy/publications/2009/geo engineering-climate/ (accessed 11 March 2012).

Sandel, Michael (1984) 'The procedural republic and the unencumbered self', *Political Theory* 12 (1): 81–96.

—— (2009) *The Reith Lectures*. Available online at www.bbc.co.uk/programmes/ b00kt7rg (accessed 31 March 2012).

Schlosberg, David and Sara Rinfret (2008) 'Ecological modernisation, American style', *Environmental Politics* 17 (2): 254–75.

Schuppert, Fabian (2011) 'Climate change mitigation and intergenerational justice', *Environmental Politics* 20 (3): 303–21.

Shearman, David and Joseph Wayne Smith (2007) *The Climate Challenge and the Failure of Democracy*. Westport CT: Praeger.

Shepard, Anna (2009) 'A case of green fatigue', *Prospect*, November, available at www.prospectmagazine.co.uk/magazine/a-case-of-green-fatigue/ (accessed 17 July 2012).

Singer, Peter (2002) *One World: The Ethics of Globalisation*. New Haven CT and London: Yale University Press.

Skinner, Quentin (1992) 'On justice, the common good and liberty' in Mouffe, Chantal (ed.), *Dimensions of Radical Democracy*. London and New York NY: Verso.

Smith, Graham (2003) *Deliberative Democracy and the Environment*. London and New York NY: Routledge.

Smith, Mark, J. and Piya Pangsapa (2008) *Environment and Citizenship: Integrating Justice, Responsibility and Civic Engagement*. London and New York NY: Zed Books.

Soper, Kate (1995) *What Is Nature?* Oxford and Malden: Wiley-Blackwell.

—— (2000) 'Future culture: realism, humanism and the politics of nature', *Radical Philosophy* 120.

Stern, Nicholas (2006) 'Stern Review: the economics of climate change' (executive summary), available at www.hm-treasury.gov.uk/sternreview_index.htm.

—— (2009) 'The rich the poor and the planet', LSE Connect, available at www2.lse.ac.uk/GranthamInstitute/Media/Articles/rich-poor-planet-NY-stern.pdf (accessed 23 July 2012).

—— (2009) *A Blueprint for a Safer Planet*. London: Vintage.

Swyngedouw, Erik (2009) 'The antinomies of the postpolitical city: in search of a democratic politics of environmental production', *International Journal of Urban and Regional Research* 33 (3): 601–20.

—— (2010) 'Apocaypse forever? Post-political populism and the spectre of climate change', *Theory, Culture and Society* 27 (2–3): 213–32.

Tambakaki, Paulina (2009) 'When does politics happen?', *Parallax* 52.

Taylor, Bob Pepperman (1996) 'Democracy and environmental ethics' in Lafferty, William M. and James Meadowcroft (eds), *Democracy and the*

Environment. Cheltenham and Lyme CT: Edward Elgar.

Terry, Geraldine (2009) 'No climate justice without gender justice', *Gender and Development* 17 (1): 5–18.

The Age of Stupid (2009) Film directed by Franny Armstrong, Spanner Films, UK.

The Day after Tomorrow (2004) Film directed by Roland Emmerich, Twentieth Century Fox.

Tietenberg, T. 'The tradable permits approach to protecting the commons', *Oxford Review of Economic Policy* 19 (3): 400–19.

Titmuss, Richard (1970) *The Gift Relationship*. London: George Allen and Unwin Ltd.

Torgerson, Douglas (1999) *The Promise of Green Politics: Environmentalism and the Public Sphere*. Durham NC and London: Duke University Press.

—— (2000) 'Farewell to the Green movement? Political action and the Green public sphere', *Environmental Politics* 9 (4): 1–19.

Torgerson, Douglas (2006) 'Expanding the green public sphere: post colonial connections', *Environmental Politics* 15 (5): 713–30.

Toynbee, Polly (2009) 'Gutless, yes. But the planet's future is no priority of ours', *The Guardian*, 19 December, 35.

Tully, James (1989) 'Wittgenstein and political philosophy' in *Political Theory* 17 (2): 172–204.

—— (2001) 'An ecological ethics for the present' in Gleeson, Brendan and Nicholas Low (eds), *Governing for the Environment: Global Problems, Ethics and Democracy*. Basingstoke: Palgrave.

Vidal, John (2009) 'Lifting the lid on climate talks', *The Guardian*, 7 November, available at www.guardian.co.uk/environment/2009/nov/07/climate-change-talks-2009 (accessed 26 July 2012).

Wara, M. (2007) 'Is the global carbon market working?', *Nature* 445: 595–6.

White, Stuart (2010) 'The future left: red, green and republican', *Red Pepper*, February 2010, available at www.redpepper.org.uk/the-future-left-red-green-and/ (accessed 16 May 2012).

Wilby, Peter (2011) 'Without talk of the common good, Earth will have to fry', *The Guardian*, 3 December, available at www.guardian.co.uk/comment isfree/2011/dec/02/george-osborne-planet-durban-failure (accessed 28 May 2012).

World Resources Institute (2005) 'Navigating the numbers', available at http://pdf.wri.org/navigating_numbers.pdf (accessed 25 July 2012).

Wynne, Brian (2010) 'Strange weather, again: climate science as political art', *Theory, Culture and Society* 27 (2–3): 289–305.

Young, Iris (1989) 'Polity and group difference: a critique of the idea of universal citizenship', *Ethics* 99 (2): 250–74.

—— (1996) 'Communication and the other: beyond deliberative democracy' in Behabib, Seyla (ed.), *Democracy and Difference*. Princeton NJ: Princeton University Press.

—— (2005) *On Female Body Experience: 'Throwing Like a Girl' and Other Essays*. Oxford: Oxford University Press.

Yuval-Davis, Nira (1997) *Gender and Nation*. London: Sage.

Zizek, Slavoj (2009) *The Parallax View*. Cambridge MA: MIT Press.

Index

160

About Zed Books

Zed Books is a critical and dynamic publisher, committed to
increasing awareness of important international issues and
to promoting diversity, alternative voices and progressive
social change. We publish on politics, development, gender,
the environment and economics for a global audience
of students, academics, activist and general readers. Run
as a co-operative, we aim to operate in an ethical and
environmentally sustainable way.

Find out more at:

www.zedbooks.co.uk

For up-to-date news, articles, reviews and events information
visit:

http://zed-books.blogspot.com

We can also be found on **Facebook, ZNet, Twitter**
and **Library Thing**.